max science primary

Student Book 2

DISCOVERING THROUGH ENQUIRY

PAT DOWER

SERIES EDITOR: **BOB KIBBLE**

macmillan education
INTERNATIONAL CURRICULUM

Macmillan Education
4 Crinan Street
London N1 9XW
A division of Springer Nature Limited

Companies and representatives throughout the world

Student Book ISBN 978-1-380-02155-7
Digital Student Pack ISBN 978-1-380-02410-7
Student Bundle Pack ISBN 978-1-380-02412-1

First published 2019

Original design by EMC Design Ltd and Macmillan Education
Page make-up by Ink Design
Illustrated by Sean Strydom, Natalie and Tamsin Hinrichsen, Magriet Brink, Bev Victor, Deevine Design
Natalie and Tamsin Hinrichsen pp 10, 11, 14, 22, 27, 28, 31, 36, 40, 60, 61, 63, 64, 66, 67, 70, 71, 73, 75, 76, 77, 81, 82, 84, 90; Deevine Design pp 78, 87, 88, 89; Magriet Brink pp3, 4, 8, 9, 12, 21; Sean Strydom pp 18, 19, 20, 25, 29, 45, 48, 52, 53, 54, 57; Bev Victor pp 86
Cover design by Sparks Publishing Services Ltd and Macmillan Education
Cover illustration by Star Publishing Pte Ltd and Jorge Santillan c/o Beehive Illustration Agency
Picture research by Susannah Jayes

The author and publishers would like to thank the following for permission to reproduce their photographs:
Alamy/Frankie Angel p15, Alamy/David Parker p59(br); **Getty** pp41, 49(tc), 85(cr); **Getty Images**/7nuit p70, Getty Images/alejandrosoto p43(cl), Getty Images/Colin Anderson p8, Getty Images/Anca Asmarnandel/EyeEm p26(cr), Getty Images/Nicholas Ayer/EyeEm p23, Getty Images/Jordan Banks p85(tl), Getty Images/Angelica Bernal/EyeEm p6(tr), Getty Images/billnoll p37(br), Getty Images/John Bower at Apex Photos p59(tl), Getty Images/Jon Boyes p82, Getty Images/Burrard - Lucas/Barcroft Media p38, Getty Images/David Clapp p47, Getty Images/commoner28th p7(cl), Getty Images/Andy Crawford p43(cr), Getty Images/dagut p76, Getty Images/Danita Delimont p49(br), Getty Images/dszc p56(cr), Getty Images/Michael Dunning p5, Getty Images/elzuaer p69(bm), Getty Images/FactoryTh p28, Getty Images/Warren Faidley p89, Getty Images/Germanovich p34, Getty Images/Nino H. Photography p10, Getty Images/Amanda Hall/Robert Harding p55, Getty Images/hannahgleg p17(cr), Getty Images/Hemera/Roxana Gonzalez p74, Getty Images/Guy Hurlebaus p24(tr), Getty Images/Injenerker p49(tl), Getty Images/iStockphoto/Bergamont p32(jumper), Getty Images/iStockphoto/real444 p32(plastic bag), Getty Images/iStockphoto/scanrail p16(cm), Getty Images/iStockphoto/Stevo24 p32(book), Getty Images/iStockphoto/Thinkstock Images/Anson_iStock p69(bl), Getty Images/iStockphoto/Thinkstock Images/Tryfonov Ievgenii p85(br), Getty Images/iStockphoto/Thinkstock/oreshkov p37(cl), Getty Images/BrianAJackson p78, Getty Images/JGI p24(bl), Getty Images/Rune Johansen p84(cr), Getty Images/Jupiterimages p2, Getty Images/RomanKhomlyak p85(tr), Getty Images/Mark Lewis p49(tc), Getty Images/Don Maison p46, Getty Images/Clare Mansell p77, Getty Images/Jonathan Mehring p13, Getty Images/mgs p83, Getty Images/Mint Images/Frans Lanting p50(cr), Getty Images/Mint Images/Art Wolfe p56(tl), Getty Images/New Zealand Transition p7(br), Getty Images/Niclasbo p88, Getty Images/Anna Pekunova p33(tr), Getty Images/Tim Ridley p32(table), Getty Images/Michael Roberts p42(tl), Getty Images/Science Photo Library p62(tr), Getty Images/seksanwangjaisuk 44(tc), Getty Images/Tetra Images p14(tl), Getty Images/Photo by Anthony Thomas p52(cm), Getty Images/GEN UMEKITA p60, Getty Images/Steve Waters p53(tr), Getty Images/WestEnd61 p6(br), Getty Images/World Perspectives p58, Getty Images/wrangel p9, Getty Images/Francesca Yorke p49(tr), Getty Images/Zhang Zheng p84(br); **PhotoAlto** p27; **Photodisc**/Getty Images p72; **REX Shutterstock**/KPA/Zuma p33(cr); **Science Photo Library**/Martyn F. Chillmaid p39; **Shutterstock**/100words p80(bl), Shutterstock/Africa Studio pp14(tmr), 53(cr), Shutterstock/Beautyimage p42(cm), Shutterstock/Joe Belanger p17(tr), Shutterstock/Stephane Bidouze p80(cr), Shutterstock/Serhii Bobyk p4, Shutterstock/canadastock p80(cl), Shutterstock/Chokswatdikorn p79(tr), Shutterstock/Civdis p33(br), Shutterstock/CRS Photo 37(cr), Shutterstock/Dja65 p53(br), Shutterstock/Ewa Studio p42(bl), Shutterstock/gcatofotografia p16(cmr), Shutterstock/gresei p16(cr), Shutterstock/Samuel Kornstein p50(br), Shutterstock/kosmos111 p26(tr), Shutterstock/Jan Kratochvila p14(cl), Shutterstock/Vasin Lee p22, Shutterstock/Viktor Lesnyh p75, Shutterstock/Pickone p44(cr), Shutterstock/rangizz p52(cr), Shutterstock/Rock and Wasp p30, Shutterstock/Kyle Santee p49(bl), Shutterstock/sdecoret p69(cl), Shutterstock/sebikus p65, Shutterstock/Shebeko p37(bl), Shutterstock/smereka p42(cl), Shutterstock/topimages p43(tc), Shutterstock/vesnation p62(bl), Shutterstock/Vlad61 p79(bc), Shutterstock/Valoga p14(tc), Shutterstock/Wstockstudio p32(spoon); **Springer Nature Limited** 14(tr); © **Stockbyte Royalty Free Photos** p58(br).

Printed and bound in Brazil

2025
21

Contents

Introducing *Max Science primary*

Max Science primary is a science course designed to meet the needs of learners following the Cambridge International Primary curriculum framework for Stages 1 to 6.

Success in science learning for young international learners and their teachers is enhanced through its practical approach to the skills of scientific enquiry and scientific vocabulary. Investigation and language activities in each unit progress learners from simple to more complex to make learning and understanding science accessible and fun.

The **Student Books** guide learners through key scientific knowledge, language and skills in the Cambridge curriculum. Each unit begins with an engaging introduction followed by activities that introduce and reinforce the skills of scientific enquiry providing opportunities for learners to develop subject knowledge through investigation.

The **Workbooks** expand the key activities in each unit for use by the learners in pairs and groups in the classroom, or as individuals at home.

To further support the course, each stage has a **Journal** for use by learners at home, with the support of adults, to reinforce the knowledge and language of each topic. Teachers can review Journals at different points throughout the term to check progress at home against progress in school.

Also available are ***Max Science primary* Digital Student Books**, which offer an accompaniment to the more traditional printed Student Books. Ideal for encouraging digital literacy, these online books mirror the print course while offering extra interactive activities and assessment opportunities at home or in class.

Student Book features

Unit openers

Introduce the unit with clearly defined learning objectives, and stimulate interest with starter questions and images.

Orientation

Starter activities engage learners and help teachers assess how much the class already knows, before exploring further.

Key words

Key words for each section are clearly shown.

Interesting facts

Link science in the classroom to science in the real world.

Exploring science

Investigations and extension questions help develop skills and understanding, and enable learners to explore, test, measure, record, interpret and predict scientific questions.

Workbook icon

Workbook icons in the Student Book direct teachers and learners to the corresponding activities in the workbook.

Assessing progress

End of topic checklists allow learners to check what they have learnt so far.

Cambridge Primary curriculum framework

In the spirit of the Cambridge Primary curriculum checkpoint tests, end of unit activities ask 'what have I learnt?' to sum up where learners have got to in their learning journey so far.*

*See *also* the online versions at www.macmillaneducationeverywhere.com for downloading and printing with complete answer spaces provided.

1 Light and dark

In this unit, I am learning:

- that light comes from different objects
- about bright light and dim light
- what darkness is
- what a shadow is
- how to make shadows
- that shadows can have different shapes.

Key words

ask
compare
dark
guess
light
Sun
torch

Where does light come from?

Look at the picture. A fire gives off light when it is lit.

In a group, think of other places where light can come from.

Tell your ideas to the rest of the class.

1.1 Where does light come from?

Key words
compare
dark
light
Sun
torch

In this section, I am learning:

- that light comes from different objects
- about bright light and dim light.

Light helps us to see.

Light comes from the **Sun**.

Light can also come from other things, like this candle.

Your teacher will turn off the lights in the classroom. Can you still see well?

Would you be able to see if there were no light at all?

What could help you see in the dark?

1 Exploring where light comes from

Your teacher has put many objects around the room.

Try each object. Which objects give off light?

Show your answers in your Workbook. WB 1

Light makes things look bright

The children's faces look bright. The light is coming from the cinema screen.

Light makes objects look bright.

Some things look bright because they make their own light. A **torch**, the Sun and a candle make their own light.

Some things look bright because they pass on light. They do not make their own light.

A mirror passes on light. It does not make its own light.

The mirror passes on the light from the candle.

2 Investigating further

Take the objects from Activity 1 into a **dark** place. There should be no light.

Which objects make their own light?

Which objects do not make their own light?

Complete the table in your Workbook.

Compare your answers with those from Activity 1.

Are they different? Why do you say so?

WB 2

Hint

When light shines onto some objects, it bounces off them.

It makes it look as if the object is giving off its own light when it is not.

Interesting fact

The Moon does not give off its own light.

The Moon looks bright when the Sun shines on it.

In the picture, you cannot see the part of the Moon that does not have the Sun shining on it.

Brighter and dimmer

Look closely at a burning candle.

Are some parts of the flame brighter than other parts?

Something that looks bright gives off a lot of light.

Something that looks dim does not give off a lot of light.

3 Looking at flames

Draw the candle flame in your Workbook. Colour it in.

Label the brighter and dimmer parts.

Describe what you see.

WB 3

Extension

One teaspoon looks brighter than the other. Why do you think that is?

WB 3

1.2 Light and dark

In this section, I am learning:

- what darkness is.

Key words

guess
shade
tell

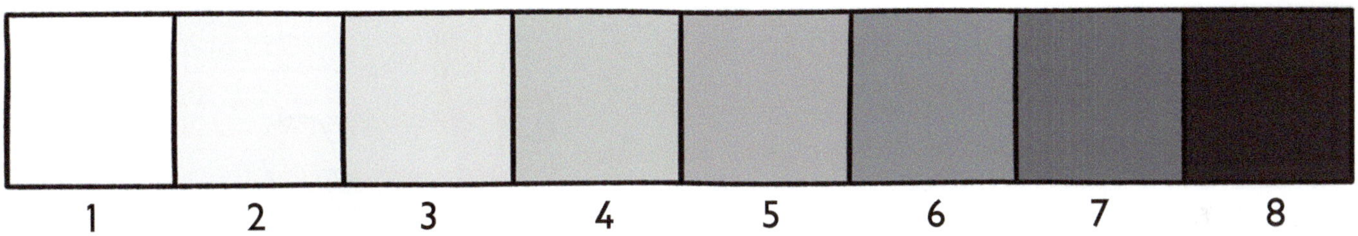

Look at the different shades in the boxes and the picture.

Use the numbers to say which shades you can see in the picture.

Can it be completely dark?

This picture was taken at night.
We say it is dark at night.

But there is often dim light at night. This means it is not completely dark at night.

It is completely dark only when there is no light at all.

Some places underground do not have any light.

A cave under the ground can be completely dark.

A torch will help you to see.

1 **Looking for complete darkness**

Look at the picture.

There is light coming from the head torch.

What happens if the head torch is switched off?

WB 4

2 **The dark box**

Your teacher will show you a dark box in the classroom.

It is completely dark inside.

Guess which of these objects you will be able to see inside the box.

Tell your partner. Show your answers in your Workbook.

Now put the different objects in the dark box. Which objects can you see?

Were your guesses right?

WB 4

Hint

Remember, some objects give off their own light.

Some objects only pass light on from elsewhere.

Extension

Martha uses a candle to help her to read in the dark.

Martha has a mirror.

She can see the real candle and the candle in the mirror.

This means that Martha can see two candles.

Does this make the room brighter than with only the real candle?

WB 5

3 Can you see a white rabbit in a very dark room?

Snowy is a white rabbit.

Trace Snowy's shape from your Workbook. Use white paper. Cut out the shape.

Find a small room or a cupboard. It must be a very dark place.

Turn on a light.

Fix the paper rabbit to the wall.

If you turn the light off, it will be very, very dark. Can you see Snowy?

Can you see anything? Why?

WB 5

This Mexican cavefish has no eyes. It lives in deep underwater caves.

Why does it not have eyes?

Checklist

Complete these statements on page 6 of your Workbook.

- I can name things that give off their own light.
- I know some objects only pass on light from elsewhere.
- I can describe what "dark" means.
- I can describe the difference between bright light, dim light and darkness.

1.3 Shadows

In this section, I am learning:

- what a shadow is
- how to make shadows
- that shadows can have different shapes.

Key words
ask
shadow

What do you see in this picture?

1 Amar's shadow

Amar has lost his **shadow**. Draw Amar's shadow for him in your Workbook.

Ask to see your partner's drawing. Do your drawings look the same?

WB 7

2 Looking at my shadow

Go outside on a sunny day. Look at your shadow.

Is your shadow completely dark?

Can you see your face in your shadow?

Look back at your drawing of Amar's shadow.
Do you now want to draw Amar's shadow differently?

WB 7

Extension

Can you see your shadow on a cloudy day?

WB 8

3 What makes a shadow?

In a dark classroom, shine a torch at the wall. How can you make a shadow?

WB 9

4 Helping Eva to understand shadows

Eva is thinking about what she learnt about shadows:

- My shadow is black.
- I have a shadow when it is completely dark.
- I have a shadow when light is shining.
- I can make a shadow by putting something in the way of a shining light.
- All objects have the same shadow.

Help Eva to say which sentences are correct. Talk about the sentences in a group.

In your Workbook, make a tick next to a sentence that is correct.

Make a cross next to a sentence that is wrong.

WB 9

Hint

Light cannot travel through a piece of wood. The wood blocks the light. There is a shadow on the other side.

What shapes can you make with shadows?

Watch the video that your teacher shows you.
Count how many kinds of animal you see.

Interesting fact

Sometimes the Moon blocks the light shining from the Sun. This casts a shadow on the Earth. We call this a solar eclipse. A solar eclipse makes it look dark during daytime.

Checklist

Complete these statements on page 10 of your Workbook.

- I know what a shadow is.
- I can draw shadows correctly.
- I can say what makes shadows.

What have I learnt?

Answer these questions on the worksheet your teacher gives you.

1. Which of these objects make their own light?

2. Why do some objects look bright even if they do not make their own light?

3. Complete these sentences:
   - It is dark when there is no _ _ _ _ _.
   - When it is completely dark, you cannot _ _ _ anything.
   - A _ _ _ _ _ _ forms when something gets in the way of light.

4. Which shadow is not correct?

2 Electricity

In this unit, I am learning:

- that many objects need electricity to work
- that circuits need a battery for them to work
- how to build an electrical circuit
- how to make bulbs light up in a circuit
- to predict whether or not a circuit will work
- about how electricity flows in a circuit
- where to find switches
- how to use a switch
- how a switch works
- about using electricity safely.

Key words

battery
bulb
buzzer
circuit
connector
switch
wire

The picture shows shining light bulbs.
What do you think makes the light bulbs shine?
Do you think the light bulbs always shine like this?

2.1 What does a battery do?

In this section, I am learning:

- that many objects need electricity to work
- that circuits need a battery for them to work
- how to build an electrical circuit
- how to make bulbs light up in a circuit.

Key words
battery, batteries
battery holder
bulb
cell
circuit
connector
electricity
filament
flow
wire

Batteries give electricity

Many things we use every day need **electricity** to work.

We can carry some of these things around with us. We say they are portable.

We can use **batteries** to make things like portable radios or a torch work.

1 Which objects need a battery?

Your teacher will show you some objects, like the ones in the pictures.

Do these objects need batteries to make them work?

WB 12

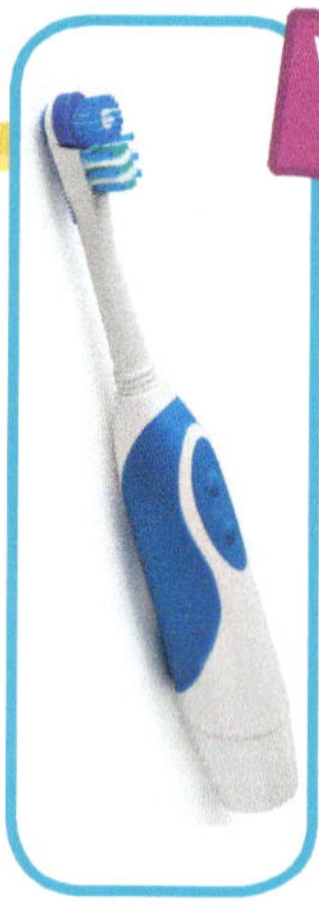

Your teacher will take a torch apart to show you some of its parts.

Guess why the bulb does not light up after the torch has been taken apart.

Interesting fact

Some batteries are much bigger than the ones we use in a torch. This picture shows a **battery** in a car.

2 Looking at batteries

Look at a battery like the ones in the picture.

Your teacher will also show you some other batteries.

Compare the different batteries. Then answer the questions in your Workbook.

WB 13

Learning about light bulbs

A light **bulb** is made of glass. There is a very thin wire inside the glass bulb.

This wire is called a **filament**.

The electricity **flows** along the filament and makes it glow. This makes the bulb light up.

3 Looking at light bulbs and bulb holders

Use a hand lens to inspect some light bulbs.

The bulb can be screwed into a bulb holder.

Describe what you see inside the bulb.

Draw the bulb in your Workbook.

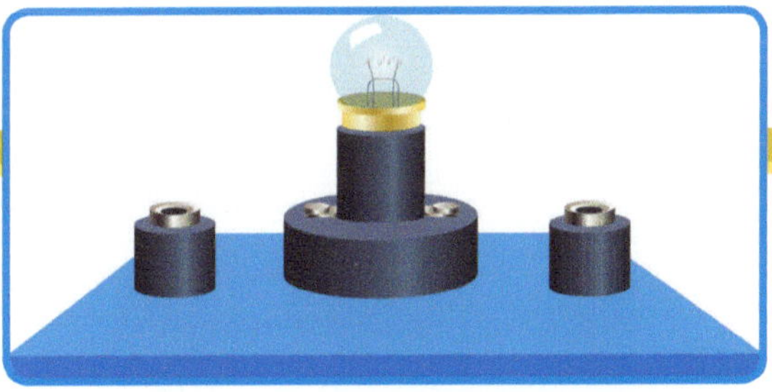

WB 14

Extension

Why do you think the filament in a bulb can break?

Complete filament

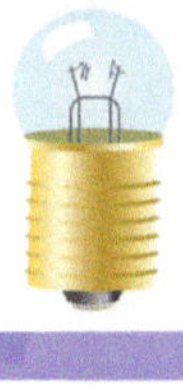

Broken filament

WB 14

Electricity flow

Electricity needs a path to flow. This path is called a **circuit**.

Look at the circuit in the picture. Electricity flows through the **wires**, the bulb and the battery all the time. We cannot see electricity.

This circuit is complete. This means there are no breaks.

If there is a break in the circuit, the electricity cannot flow. The bulb will not light up.

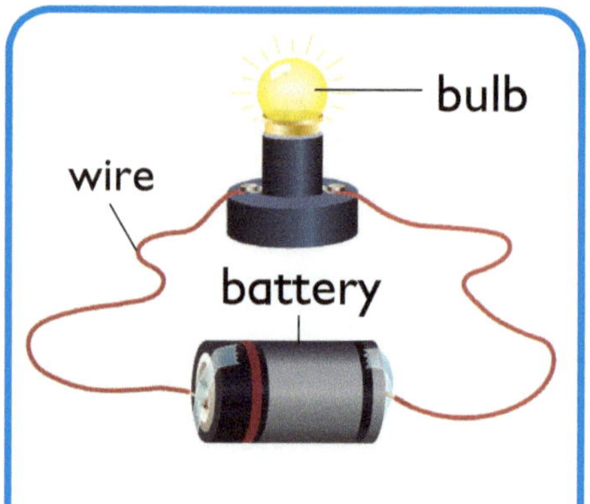

A circuit

Hint

Scientists use the word "**cell**" when they talk about a single battery like the one in the picture. In everyday language we usually just say "battery".

Making a circuit

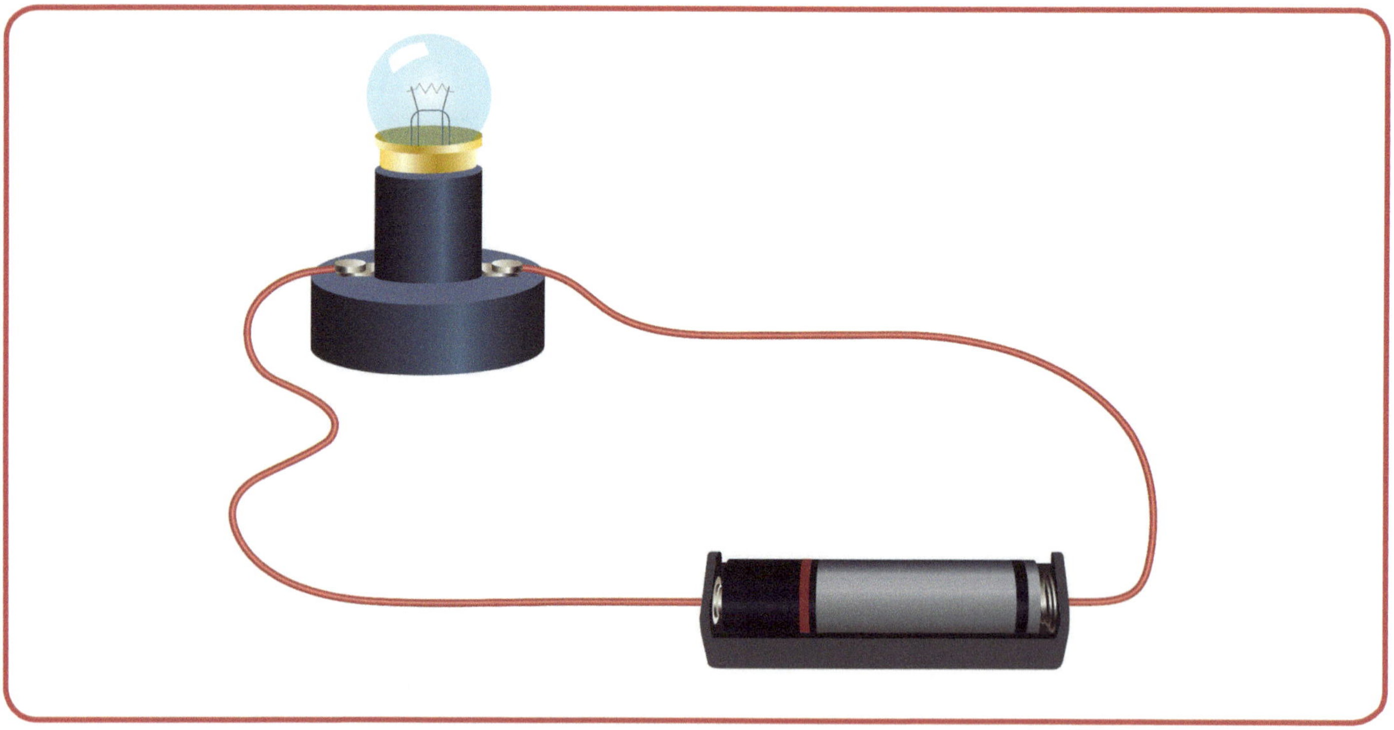

We can use a battery, two **connector** wires and a bulb to make a circuit.

We can place the bulb in a bulb holder. The bulb holder makes it easier to connect the wires to the bulb.

A battery can also be in a **battery holder**. This makes it easier to use.

4 Completing a circuit

The first picture shows a bulb in a holder. The second picture shows a battery in a holder.

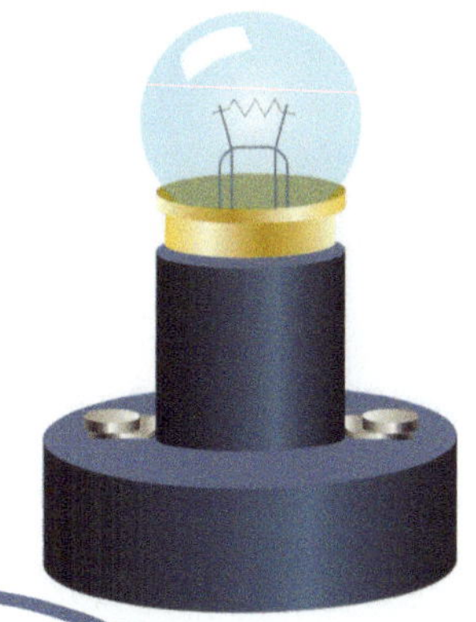

How many wires will you need to connect these items and make the bulb light up?

Draw a picture of the circuit in your Workbook.

Complete the sentences in your Workbook.

WB 15

Checklist

Complete these statements on page 16 of your Workbook.

- I can name some things that need electricity to work.
- I can say what a battery does.
- I know the parts of a circuit.
- I can build a circuit.
- I can say why a circuit must be complete for a bulb to light up.

2.2 Exploring circuits

In this section, I am learning:

- to predict whether or not a circuit will work
- about how electricity flows in a circuit.

Key words
guess
predict

Electricity needs a circuit to flow

When the wires touch the + and – on a battery, electricity can flow around the circuit.

The flow of electricity lights up the bulb.

Look at the picture. How can you make a circuit?

1 Which circuit will work?

Look at the circuits below.

Only one of these circuits will work.

Predict which circuit will light up the bulb. Then test your ideas.

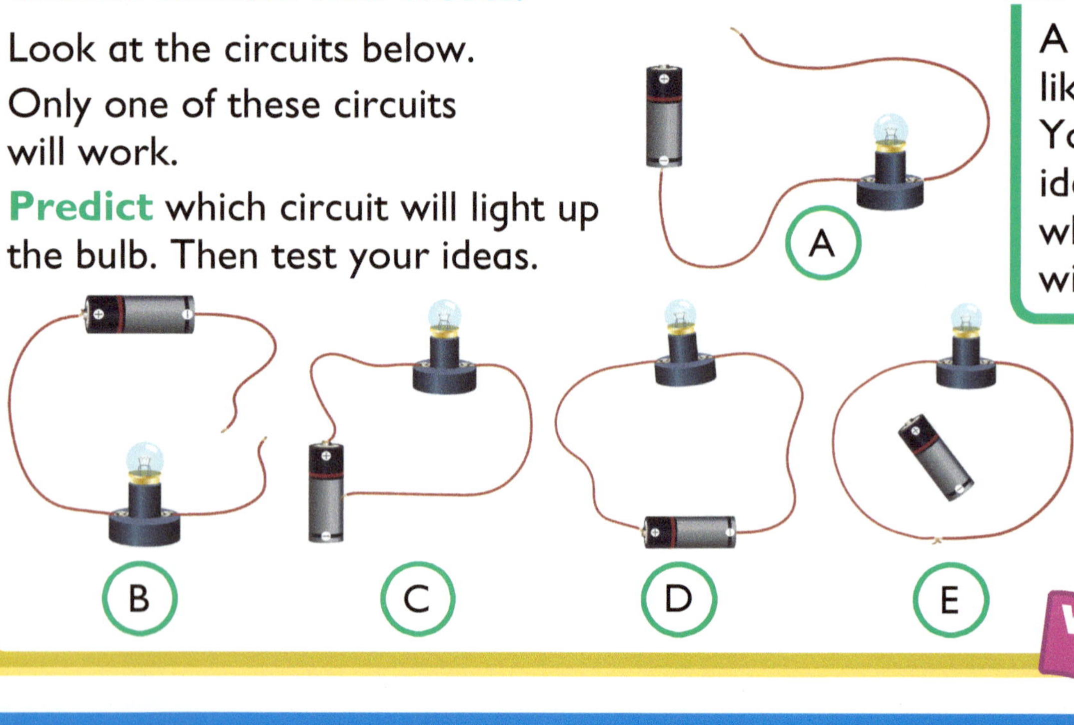

Hint

A prediction is like a **guess**. You use your ideas to say what you think will happen.

WB 17

Scientists test their ideas. This helps them to see if their ideas are correct. Testing also helps scientists to know when to change their ideas.

Test your ideas some more. What happens if you do not have a bulb holder?

Extension

Use different coloured wires to make circuits.

Does the colour of the wires make any difference?

What material is inside each wire?

WB 18

2 Testing Toby's idea

Toby has a bulb, two wires and a battery, but he does not have a bulb holder.

Toby thinks he can still make the bulb light up.

Do you agree with Toby? Explain why.

Test your idea.

WB 18

3 Making a cosmic energy ball light up

Your teacher will show you a cosmic energy ball. Try to make it light up.

Why does the ball light up? Is it magic?

Complete the sentences in your Workbook.

WB 18

Interesting fact

Lightning is very powerful electricity. It is so powerful that it can light up the sky.

4 Playing circuits

Work in a group.

You are going to be a circuit. Let one learner play the part of a battery. Let another learner play the part of a bulb.

- **Circuit 1:** Stand in a circle and hold hands so that the electricity can move around the circuit.

What happens when two people in the circuit are not holding hands?

- **Circuit 2:** Now stand in a circle and hold one loop of rope in all your hands.

What happens if one person is not touching the rope?

WB 19

Interesting fact

These decorations are made by connecting many bulbs together.

Extension

Try to draw a circuit that will make two bulbs light up at the same time.

WB 20

Checklist

Complete these statements on page 20 of your Workbook.

- I can connect a battery and bulb to make the bulb light up.
- I can predict when a circuit will make a bulb light up.
- I can test my predictions.
- I know that electricity flows in a complete circuit.

2.3 Switches and staying safe

In this section, I am learning:

- where to find switches
- how to use a switch
- how a switch works
- about using electricity safely.

Key words
break
buzzer
connect
dangerous
lamp
socket
switch

Your teacher will show you a battery alarm clock.
What are two ways of turning off the alarm?
Which is the easiest way?

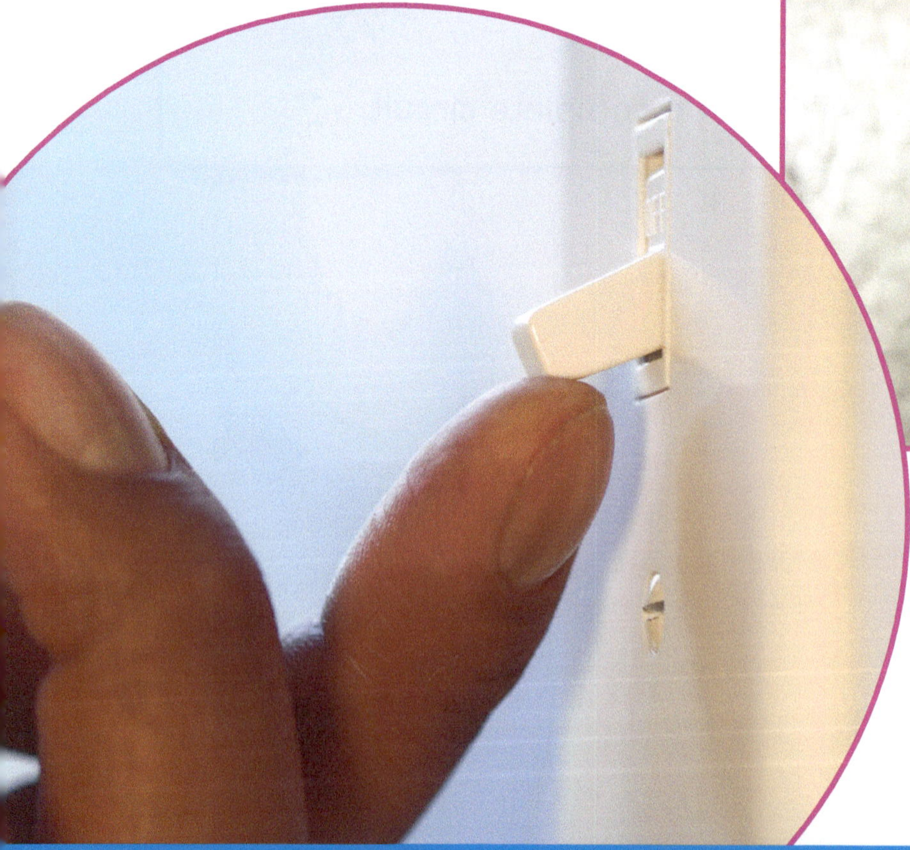

A circuit with a switch

A **switch** makes it easy to turn something on or off.

1 A switch in a circuit

Follow the steps below to **connect** a switch in a circuit. Use the switch to turn the **lamp** on or off.

Now build the same circuit with a **buzzer** instead of a lamp. Use the switch to turn the buzzer on or off.

In your Workbook, draw the circuits you built.

What does the switch do in this circuit?

WB 21

Hint

A lamp is a bulb in a bulb holder. Lamps come in different sizes and shapes.

1. Use a circuit that can make a lamp light up.

2. Take the end of one wire off the bulb holder.

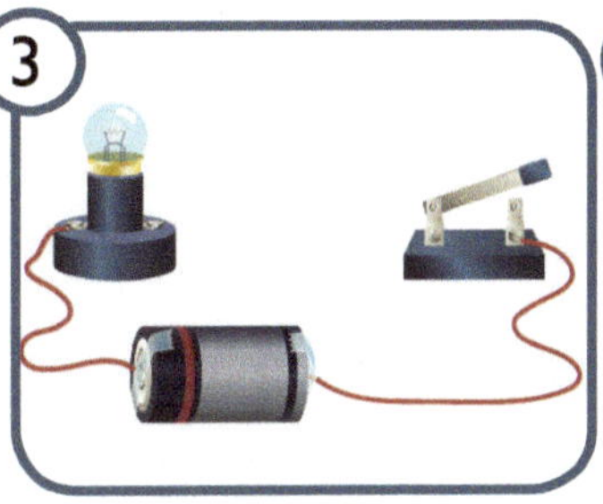

3. Connect this wire to the switch.

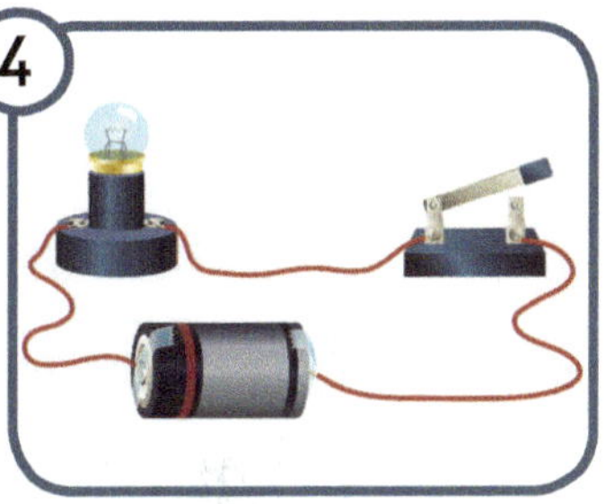

4. Use another wire to connect the other side of the switch to the bulb holder.

2 How does a switch work?

Look carefully at the switch you used in Activity 1. Also look at other switches.

Switches work by starting or stopping the flow of electricity.

How does a switch start or stop the flow of electricity?

WB 21

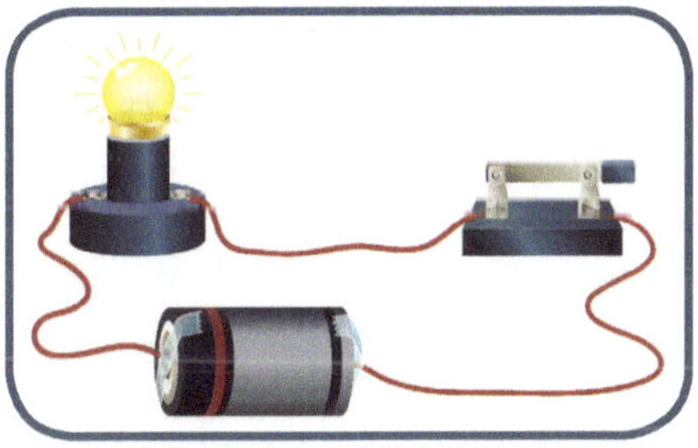

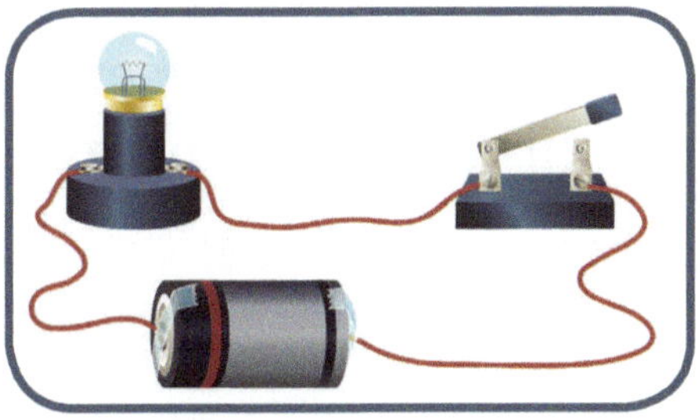

Hint

When the switch is on, electricity can flow. The circuit is complete.

When the switch is off, it makes a **break** in the circuit. The circuit is not complete. Electricity cannot flow.

Electricity in buildings

In our homes and schools, many of the things that need electricity do not have batteries. We can connect these things to a **socket** in the wall.

The electricity flows to buildings along wires underground or in the air. From here the electricity flows into wires in the walls of the building.

This kettle is connected to a socket in the wall.

3 Looking for switches and sockets

Look around your school for different switches and sockets.

What do the switches do?

What are the sockets used for?

WB 22

Hint

Electricity can be **dangerous**. Do not touch any of the switches or sockets unless your teacher says it is safe.

4 Making an alarm

Tya has been given a beautiful bracelet for her birthday.

Tya keeps the bracelet in a box when she is not wearing it.

She wants an alarm to sound if anyone opens the box.

Can you help her make an alarm using a circuit with a buzzer?

Here is an idea:

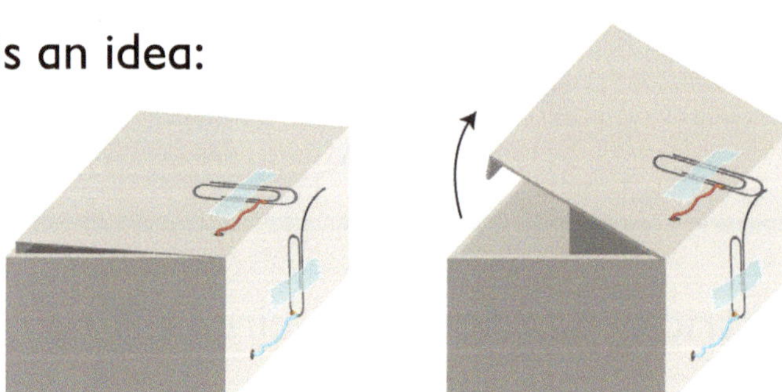

WB 22

Extension

Look at the circuits in your Workbook. For each one, say if closing the switch will turn the buzzer on.

WB 23

Staying safe around electricity

The batteries you have been using give only a small amount of electricity. This makes them safe to use. The electricity in buildings is much more powerful. It is dangerous because it can shock you. It can even kill you.

We must always be careful when working with electricity. We must also be careful when working near electricity.

These wires carry electricity to our homes.

5 A safety slogan

Look at the advice from Samir, Padma, Jared and Alena on this page and the next one.

Draw a poster about working safely with electricity.

Then add a slogan to your poster to help people stay safe.

WB 24

Hint

A slogan is a short sentence to help people remember something important.

Interesting fact

Electricity cannot pass through all materials.

It can pass easily through metal. It cannot pass through plastic.

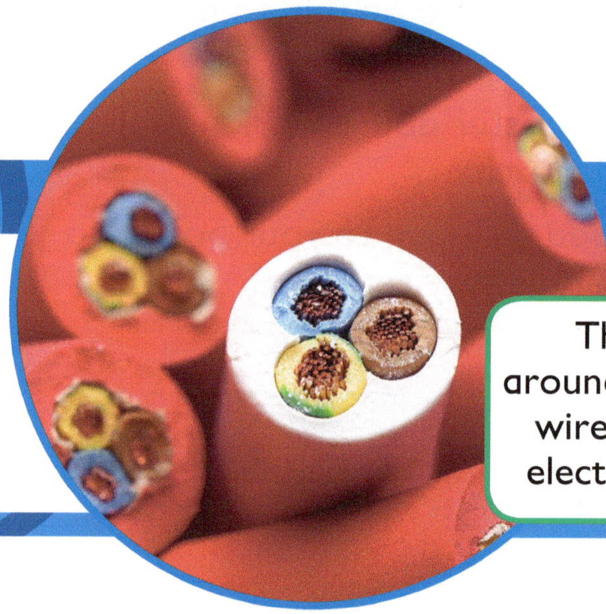

Checklist

Complete these statements on page 25 of your Workbook.

- I can find different kinds of switches.
- I can make a circuit with a switch.
- I know how a switch turns something on or off.
- I know that electricity can be dangerous.
- I know some ways to stay safe when using electricity.

What have I learnt?

Answer these questions on the worksheet your teacher gives you.

1

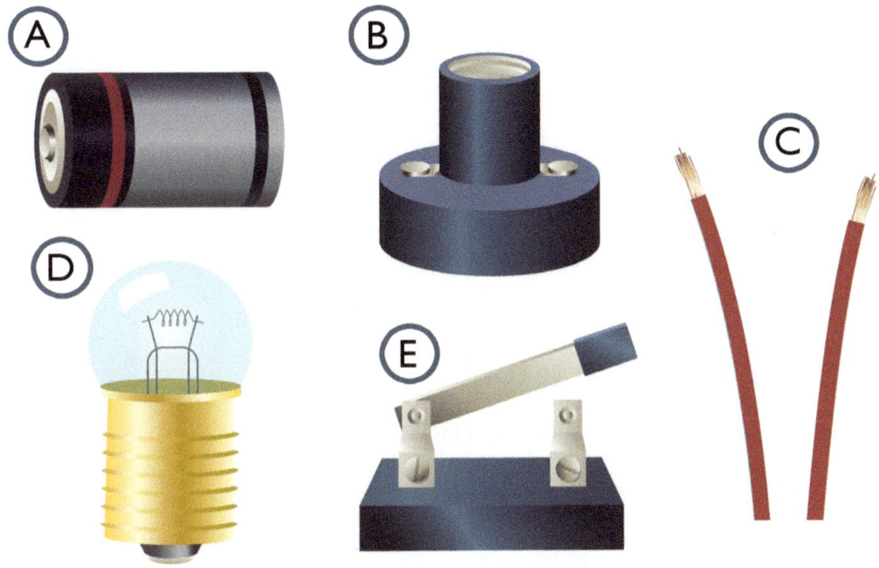

- Name the parts of a circuit shown in pictures A–E.
- What job does part E do in a circuit?

2 What can you change so that the bulb will light up in each circuit?

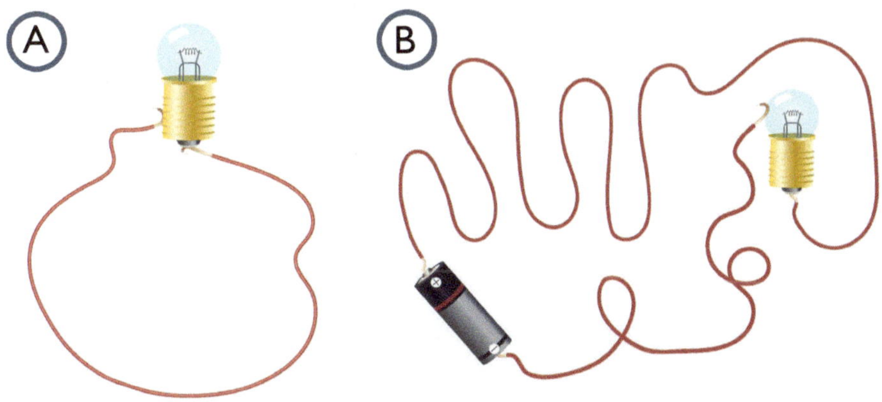

3 Changing materials

In this unit, I am learning:

- how the shape of some materials can be changed
- to predict whether a material will change shape
- how heating or cooling can make some materials change
- how water changes when it is heated or cooled
- about melting
- how to record and compare observations
- about dissolving materials
- how to record results in a table
- about natural and man-made materials.

Key words

bend
cool
heat
melt
solid
squash
stretch
twist
water droplets

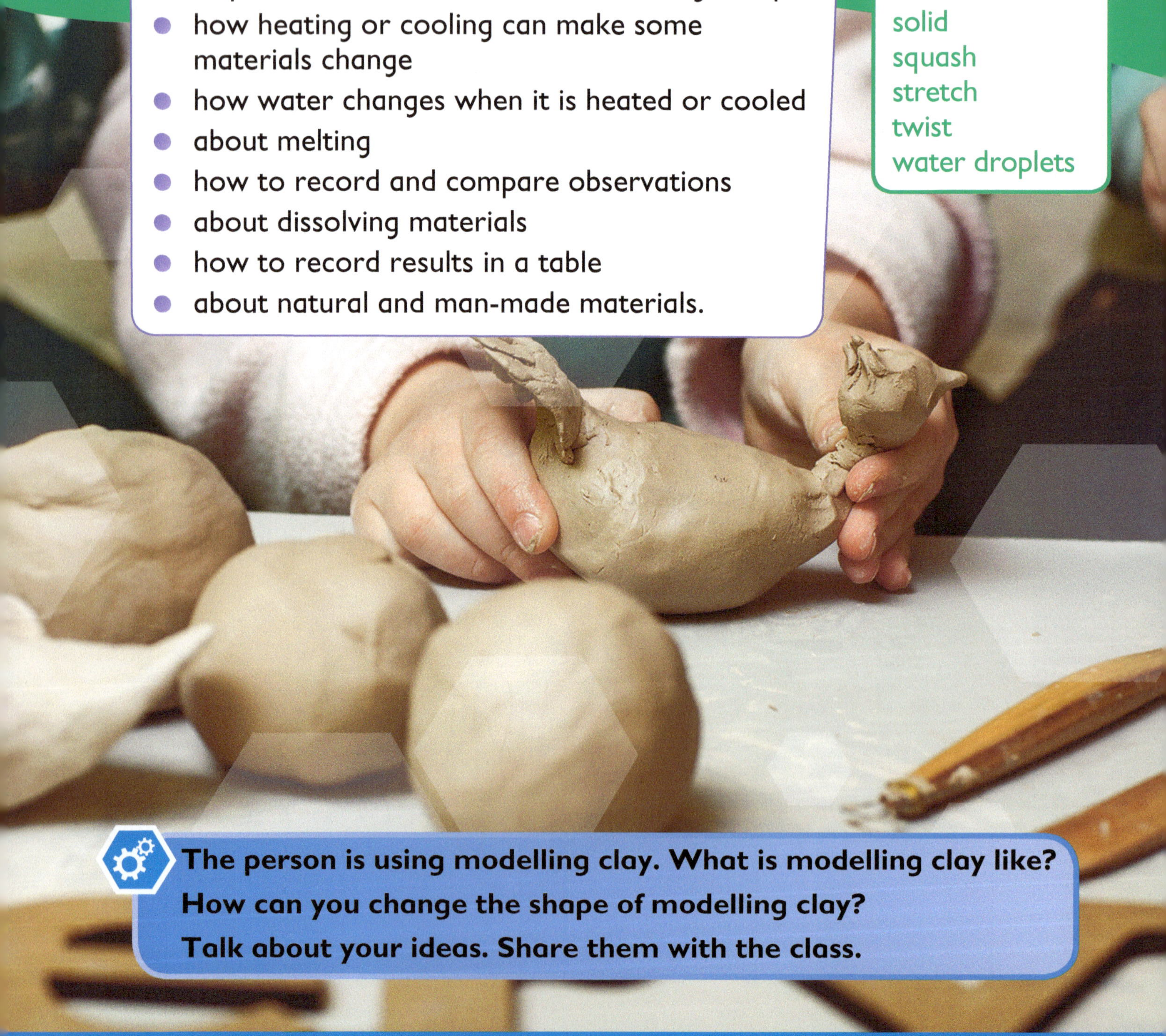

The person is using modelling clay. What is modelling clay like?
How can you change the shape of modelling clay?
Talk about your ideas. Share them with the class.

3.1 Changing shape

In this section, I am learning:

- how the shape of some materials can be changed
- to predict whether a material will change shape.

Key words

bend
change
material
predict
rigid
shape
squash
stretch
twist

Try making these different shapes with a piece of modelling clay:

- **Banana**
- **Fish**
- **Snake**
- **Cat**
- **Tree**

The shape of a material

A **material** is what something is made of. You can use a force to **change** the **shape** of some materials.

Here are some ways:

You can **twist** some materials. This uses a turning force.

You can **stretch** some materials by pulling on them.

You can **squash** some materials by pushing on them.

You can use a push or a pull to **bend** some materials.

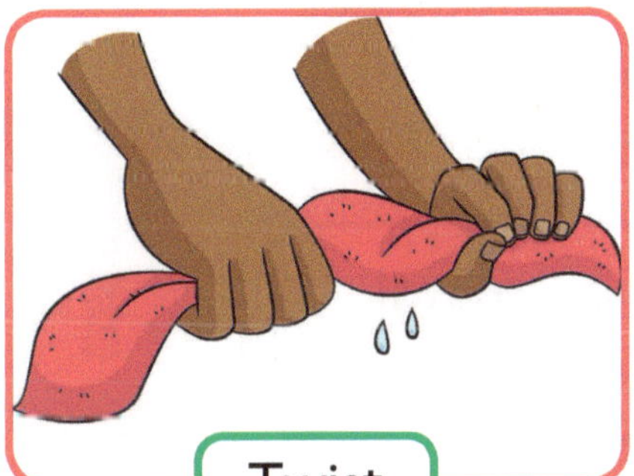

Twist

Stretch

Squash

Bend

1 Changing the shape of modelling clay

Make four balls of modelling clay.

Change the shape of each ball by:

- squashing it
- twisting it
- stretching it
- bending it.

In your Workbook, draw the new shapes you have made.

WB 27

Extension

Use modelling clay to make a model of your favourite animal.

What actions did you use?

WB 28

2 Do all materials change shape?

Look at these objects.

What material is each object made from? Choose words from the box.

wood	plastic	metal
paper	wool	

Think about some actions that can make materials change shape.

Predict whether these materials can change shape.

Now test each material to see if you were correct.

WB 29

When you stretch an elastic band it pulls back with a force.

Wood and iron are strong materials. Many strong materials are also **rigid**.

A big force is needed to change the shape of a rigid material.

Look at the bow and arrow in this picture.

What is each item made from?

Can the materials change shape?

Extension

Look at the railway track in this picture. It is built from metal. The rails do not usually look like this.

What shape do you think the track used to have?

What do you think caused the shape to change?

WB 29

Checklist

Complete these statements on page 30 of your Workbook.

- I can say how a material's shape can change.
- I can predict whether a material's shape will change.
- I can test my predictions.

Interesting fact

This fishing rod is made from a material called carbon fibre. You can change the shape of a carbon fibre fishing rod by bending it. The rod changes back to being straight after you have stopped bending it.

3.2 Can heating or cooling change materials?

In this section, I am learning:

- how heating or cooling can make some materials change
- how water changes when it is heated or cooled
- about melting
- how to record and compare observations.

Key words

cool
heat
irreversible
melt
reversible
solid
water droplets

Look at the ice in this picture.
What is happening to the ice?
Why does this happen?
Talk about it in your group.

1 Cooling materials

Your teacher will give you some materials.

What is each material like? Write your answers in your Workbook.

- Now put a small amount of each material into a **cool** place, like a fridge.
- Also put a small amount of each material into a very cold place, such as a freezer.

After a few hours, look at the materials again. What has happened?

Do they look or feel the same as before?

Fill in the table in your Workbook.

WB 31

Solid materials

A **solid** material is usually hard. A solid material has a fixed shape.

Water is not a solid. Water is not hard. Water has no shape of its own.

But when water freezes, it becomes solid. Ice is a solid.

Changes

Materials can change when they warm up.
Materials can change when they get cold.

In some materials, the change is **reversible**. This means that the material can change back to how it was *before* it got warm or cold.

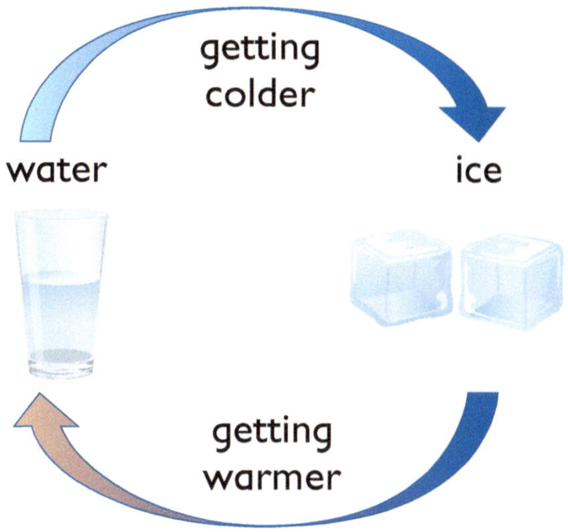

Reversible changes

2 Predicting changes in materials

- Think about what the materials in Activity 1 were like before they were cooled.
- Then think about what they were like when they got very cold.

Predict what will happen when the materials warm up again.

Will they be the same as they were before?

WB 33

3 What happens when you heat water?

Your teacher will start to **heat** cold water in a container.

Look carefully. What changes do you see?

WB 33

When water is heated, some of the water escapes into the air. We can see a fine mist. This mist is made up of tiny **water droplets**.

4 Collecting tiny water droplets from the air

In Activity 3, you saw tiny water droplets in the air. Can we collect the droplets from the air again?

Watch what your teacher does.

In your Workbook, draw what you can see on the tile.

Tell your partner why you think this happens.

WB 34

Some solid materials can **melt** when they are heated.

Butter can melt when it is left outside in the Sun.

Ice can melt when it is left outside in the Sun.

Hint

Heating and cooling water can cause reversible changes.

5 Investigating melting

It is nearly lunchtime and Kate is hungry.

She wants to have soup for lunch. But the soup is frozen. What can she do?

She asked Amir, Jessica and Xuan for some advice.

In your group, talk about the children's ideas. What will work best?

Do you have other ideas?

Your teacher will give you cubes of frozen soup so that you can try it yourself.

How well did each idea work?

WB 35

Extension

Tom has cooked his soup. But now it is too hot to eat.

What can Tom do to cool his soup?

WB 35

Materials do not all change in the same way when they get hot.

Sometimes the changes are **irreversible**. This means that the material cannot change back to how it was before it was heated.

6 How do materials change when they are baked?

Look at some modelling clay and dough before they are baked in an oven.

Touch them and squeeze them. Can you change their shape?

Then look at some modelling clay and dough that have been baked in an oven.

- What was each one like before being baked?
- What is each one like after being baked?
- Do the clay and dough look and feel the same after being baked?

Complete the table in your Workbook.

WB 36

Modelling clay before baking

Modelling clay after baking

Bread dough before baking

Bread dough after baking

Materials can change when they are heated. Food changes when it is cooked.

Clay changes when it is baked. Baked clay can last a very long time. These sculptures from China are more than 2500 years old.

What happens when you heat these foods in boiling water?

- **Rice**
- **Carrots**
- **An egg**
- **Pasta**

Extension

Does baking cause a reversible change?

- Can baked modelling clay change back into soft clay?
- Can baked dough change back into unbaked dough?

WB 37

Checklist

Complete these statements on page 37 of your Workbook.

- I know that heating or cooling a material can make it change.
- I can say how water changes when it is heated or cooled.
- I know that some solid materials can melt when they get hot.
- I can investigate how to make a material melt fast or slowly.
- I can record and compare my observations.

3.3 Dissolving

In this section, I am learning:

- about dissolving materials
- how to record results in a table.

Key words

compare
dissolve
measure
results table

Your teacher will stir some sugar into water.

What has happened to the sugar?

Mixing materials in water

When some materials are mixed into water, they **dissolve**. This means that you cannot see the material anymore after you mix it into the water.

Not all materials can dissolve in water. You can still see them after you mix them into the water.

1 Sugar and water

Place a spoonful of sugar into a glass of water. Do not stir.

Can you see the sugar grains?

Now stir the water. Watch what happens.

Predict what you think the water will taste like.

WB 38

Hint

Never taste anything in Science class unless your teacher says it is safe.

2 Which materials dissolve?

Your teacher will give your group some materials.

Mix each material into water.

Which materials dissolve?

WB 39

Using a **results table** is a good way to record what you find out. A results table has headings.

Hint

If the water looks cloudy, it is because the material has not dissolved in the water.

3 Results table

Record your results from Activity 2 in the table in your Workbook.

Write two sentences about dissolving.

WB 39

4 Does a sugar lump dissolve as easily as sugar granules?

Place a sugar cube in a glass of water. Stir the water gently for five counts.

Place a spoonful of sugar granules in another glass of water. Stir the water gently for five counts.

Compare how well each type of sugar dissolves. Show the answers in your Workbook.

WB 40

Extension

Do you think a material dissolves faster in cold water or in warm water?

Test your idea. **Measure** how long it takes for a material to dissolve.

Write your answers in your Workbook.

WB 40

Interesting fact

There is a lot of salt dissolved in seawater. Sharks can drink salty seawater because their bodies have special systems to cope with it.

Checklist

Complete these statements on page 41 of your Workbook.

- I can say what happens when a material dissolves.
- I can investigate whether a material dissolves or not.
- I can record my results in a table.
- I can name some materials that dissolve.

3.4 Natural and man-made materials

In this section, I am learning:

- about natural and man-made materials.

Key words
man-made
natural

Look at the picture.
What materials do you see?
Where do materials come from?

Some materials are **natural**.

Some materials are **man-made**.

Natural materials

Natural materials come from plants, animals or the ground.

Diamond is a natural material. A diamond comes from the ground.

Wood is a natural material. Wood comes from trees.

Leather is a natural material. Leather comes from the skin of animals.

1 Where do these materials come from?

Work with a partner.

Look at the natural materials your teacher gives you.

Do they come from a plant, an animal or the ground?

WB 42

Man-made materials

Man-made materials are not found in nature. People change natural materials in some way to make man-made materials.

Steel is a man-made material. It is made by melting iron and mixing it with carbon.

Plastic is a man-made material. It is made from crude oil.

Glass is a man-made material. It is made by melting sand.

2 Uses of some man-made materials

Look at the materials in the pictures above. What other objects are made from these materials?

Your teacher will show you some more man-made materials. What are the materials used for?

WB 42

3 Natural and man-made materials

Look for natural and man-made materials around your school.

Draw a map of your school. Label the places.

Where did you find natural materials? Make a green "N" at these places.

Where did you find man-made materials? Make a black "M" at these places.

Look at Fabian's map as an example.

WB 43

4 Natural and man-made materials on display

As a class, make two large posters to show:

- natural materials
- man-made materials.

WB 43

Hint

When objects are made from natural materials, you can still see what the natural material is.

Extension

This spoon is man-made, but the material is natural. The spoon is made from wood.

Name two more objects made from natural materials.

What has been done to the materials to make each object?

WB 44

Plastic is a useful man-made material.

What objects would not exist without plastic?

Interesting fact

Crude oil is used to make many materials. Plastic, polythene, petrol, diesel and tar all come from crude oil.

The surface of this road is made from tar.

Checklist

Complete these statements on page 44 of your Workbook.

- I can name some natural materials.
- I can say what “man-made” means.
- I can name some man-made materials.

What have I learnt?

Answer these questions on the worksheet your teacher gives you.

1. Which actions were used to change the shape of the material? Choose the correct word for each picture.

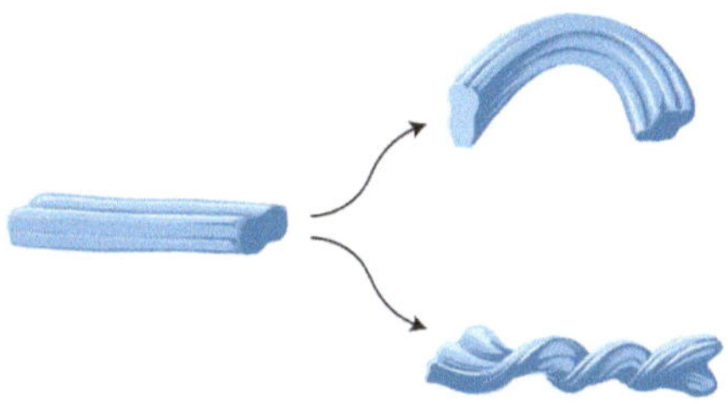

bending	stretching
twisting	squashing

2. Choose the right word to complete these sentences:
   - When ice melts it becomes **hard** / **runny**.
   - Ice melts faster when it is **hot** / **cold** outside.
   - Baking bread dough makes the dough **stretchy** / **firm**.

3. Sort these materials into two groups:
   - materials that dissolve
   - materials that do not dissolve.

clay flour salt sand stones sugar wood

4. Which sentences are true?
   - Wood is a man-made material.
   - Wool is a natural material.
   - Man-made materials are made by a man but not a woman.
   - All natural materials come from animals.
   - Plastic is a man-made material.

4 Looking at rocks

In this unit, I am learning:

- about different types of rock
- about the difference between pebbles, sand and soil
- to test how hard a rock is
- about one way that rocks are formed
- about the difference between natural and man-made building materials
- how to build a strong wall
- how natural and man-made building materials are used.

Key words

group
man-made
natural
rock
sandy
soil
stone

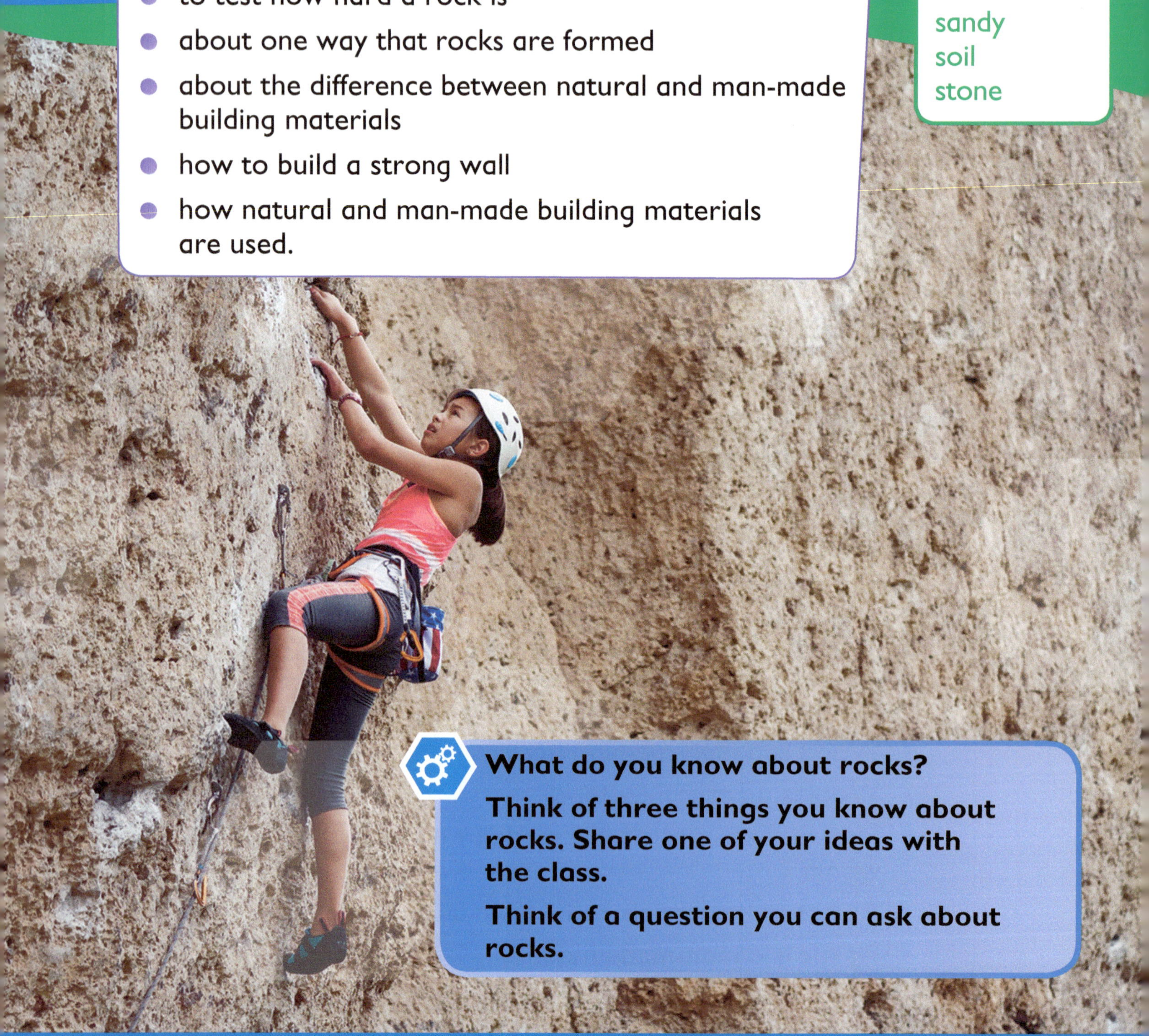

What do you know about rocks?

Think of three things you know about rocks. Share one of your ideas with the class.

Think of a question you can ask about rocks.

4.1 Different types of rock

In this section, I am learning:

- about different types of rock
- about the difference between pebbles, sand and soil
- to test how hard a rock is
- about one way that rocks are formed.

Key words

compare
group
liquid
rock
sample
sand
sandy
soil
stone
test

Rocks and stones

Rocks are solid, natural materials. **Stone** is another word for **rock**.

How many types of rock can you see in the picture?

1 Hunting for rocks

Go outside and look for pieces of different types of rock or stone.

How many different types can you find?

Draw some of them in your Workbook.

WB 47

Hint

Do not damage any rocks that you find.

Rocks can be different colours, shapes and sizes.

2 **Looking at and describing rocks**

Work with a partner. Look carefully at the rocks you collected earlier.

Describe one of the rocks to your partner without showing it to them.

Ask your partner to guess which rock you described. Let them draw a picture of it in their Workbook.

WB 47

3 **Sorting rocks**

Look at the rocks you found earlier.

Think about the features of the rocks. Are they:

- hard or soft
- light or dark
- shiny or dull?

Which other features can you describe?

Choose one pair of features to sort your rocks into two **groups**.

WB 48

Pebbles, sand and soil

Big rocks get broken up over many years. Pebbles are smaller pieces of stone that have been smoothed.

Sand forms when rock is ground down.

Soil is a mixture of very small pieces of rock and other materials. Sometimes soil feels quite **sandy**.

4 Comparing pebbles, sand and soil

Look closely at some pebbles, sand and soil. Use a magnifying glass.

Rub the pebbles between your fingers.

Do the same with the sand and then the soil.

Compare each material. Write down some of their features in your Workbook.

Pebbles

Sand

Soil

WB 48

Comparing how hard rocks are

Some rocks are hard. It is difficult to scratch or break them.

Granite is a very hard rock.

Other rocks are less hard. It is possible to break bits off them.

This sculpture is made of sandstone. Sandstone is not as hard as granite.

5 Testing the hardness of rocks

Your teacher will give you pieces of two different rocks.

Draw the two rocks in your Workbook.

Compare the rocks:

- What do the surfaces look like?
- What do the surfaces feel like?

Rub the two rocks together. Do any bits break off the rock?

WB 49

Extension

Your teacher will give you **samples** of different rocks. You will **test** the hardness of each rock by scratching it with a piece of metal.

- Give each rock a number.
- List the rocks in order from the hardest to least hard.

WB 49

How do new rocks form?

The Earth is made of rock. Some rocks we see come from inside the Earth.

It is so hot inside the Earth that the rock melts. When something melts, it becomes runny. We say it is **liquid**.

When a volcano erupts, the liquid rock flows out onto the surface of the Earth.

When the molten rock cools down, it becomes solid. It can look like this.

The oldest rocks on Earth are over 4000 million years old.

6 Volcanoes

Watch a video clip showing how liquid rock flows from a volcano.

What is the rock like when it flows from the volcano?

What is the rock like when it has cooled down?

WB 50

Do all solids melt if they get hot enough?

Interesting fact

Volcanoes even erupt underneath the sea.

When the liquid rock turns into solid rock, it can form new islands.

Checklist

Complete these statements on page 50 of your Workbook.

- I can compare different types of rock.
- I can compare pebbles, sand and soil.
- I can test how hard a rock is.
- I know of one way that new rocks can form.

4.2 Natural and man-made building materials

In this section, I am learning:

- about the difference between natural and man-made building materials
- how to build a strong wall
- how natural and man-made building materials are used.

Key words
bonding material
building material
man-made
natural

Natural building materials

We can use stone as a **building material**. Stone is a **natural** material.

A dry-stone wall

A brick wall

Look at the walls in these pictures. What differences can you see?

1 Building a pebble wall

Try to build a small dry-stone wall like the one in the picture.

Use pebbles instead of big stones.

Look for pebbles that fit together to build the wall.

Use other small pieces of stone to put in the gaps.

Draw the wall you built. Describe the wall in your Workbook.

WB 51

Interesting fact

The Great Wall of China is the longest wall in the world. It is about 21 000 km long and is more than 2300 years old.

2 Building a stone wall with bonding material

Build another wall using pebbles. This time use a **bonding material** to hold the pebbles together. You can use modelling clay as a bonding material.

In your Workbook, draw four pictures to show the stages in building your wall.

How strong is your wall now?

WB 52

Man-made building materials

Bricks are a **man-made** building material. Bricks are made by baking clay.

Clay is soft. When clay is baked, it becomes hard.

Clay bricks are rectangular. We can build straight walls with bricks.

We can use bricks to build houses and make pavements.

3 Making bricks

Make clay bricks as shown in the pictures.

Try to make your bricks all the same shape and size.

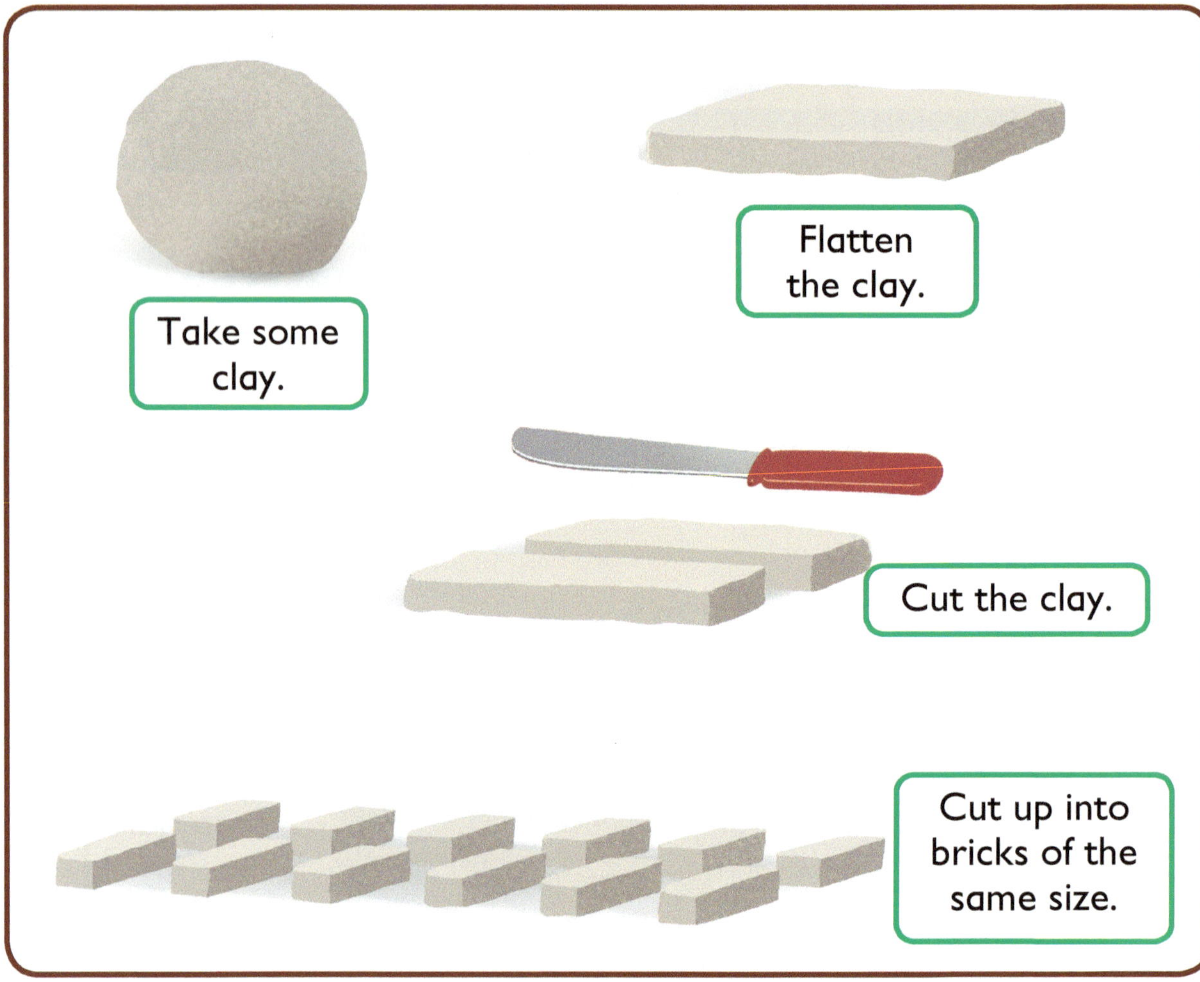

Give the bricks to your teacher to bake.

- What is the clay like before it is baked?
- What do you think the bricks will be like once the clay is baked?

Write your ideas in your Workbook.

WB 53

4 Building a wall with bricks

There are two ways to build a wall. Look at these pictures:

overlapping bricks

bricks not overlapping

Work with your partner. Use the clay bricks you made in Activity 3 to build these two walls.

Talk about the two walls:

- Which wall is stronger?
- Which wall was easier to build?
- How can you make the walls even stronger?

Draw and write about the wall you built in your Workbook.

WB 53

Hint

A wall is stronger if the bricks overlap than if they do not overlap.

Extension

Think about why we sometimes use natural building materials.

Think about why we sometimes use man-made building materials.

WB 54

Interesting fact

The Chrysler Building in New York, USA is the tallest brick building in the world. It was built in 1930.

What materials are good for making the different parts of a house?

Think about the materials for the roof and the floor.

This rock can be split into thin layers to make roof tiles. This rock is called slate.

This rock has been cut and polished. This rock is called marble.

Checklist

Complete these statements on page 54 of your Workbook.

- I can say how to build a strong wall.
- I can name some natural and some man-made building materials.
- I can compare natural and man-made building materials.

What have I learnt?

Answer these questions on the worksheet your teacher gives you.

1
- Look at the materials in the picture. Use the numbers to sort them into two groups.
- What feature did you use to sort the materials?
- How many materials are in each group?

2 Lucy tested four rocks to see how hard they were. She scratched each one with a metal knife.
- Look at the picture of Lucy's rocks. List the rocks in order from the hardest to the softest.

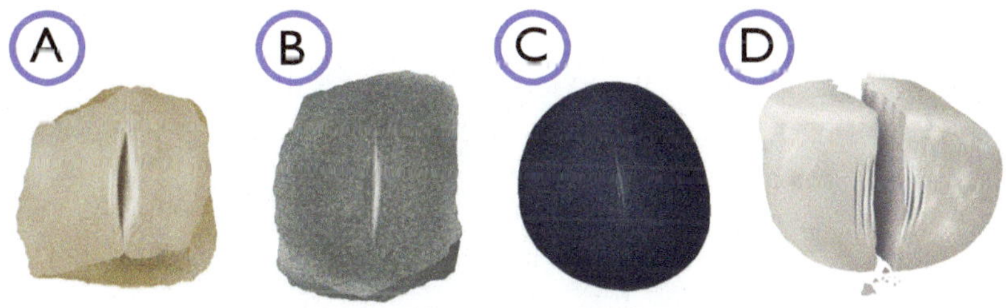

- Why is it important to use the same knife for all of the hardness tests?

3 Look at the two walls in the picture below.

- Which one should be the strongest?
- Why do you say so?

5 Day and night

In this unit, I am learning:

- how shadows change during a day
- why shadows change during a day
- that we can tell the time using the Sun's position
- how a sundial works
- how to make a sundial
- how the Earth moves on its axis
- how the movement of the Earth causes day and night
- how the Moon moves
- why we cannot always see the Moon at night.

Key words
compare
Earth
Moon
shadow
spin
Sun

Why does this photo of the Earth look different from the other one?

5.1 Looking for shadows

In this section, I am learning:

- how shadows change during a day
- why shadows change during a day.

Key words

compare
depends
equator
light
measure
position
shadow
Sun
unit

Where can you see shadows in this picture? How many shadows can you see?

Go outside on a sunny day and look for shadows.

Try to find five different shadows around you.

How do shadows form?

A **shadow** forms when something blocks the path of **light**. The light has to be bright for a shadow to form.

The **Sun** gives bright light. If you stand outside on a sunny day, you can see your own shadow. This is because your body blocks the path of the sunlight.

Changing shadows

Your shadow is not always in the same place.

Where your shadow is **depends** on where the light is shining from.

- If the light is in front of you, your shadow will be behind you.
- If the light is behind you, your shadow will be in front of you.

We can look at the **position** of a shadow to know where the Sun is.

The girl's shadow is in front of her. This means the Sun is behind her.

1 Where is the Sun?

Look at the picture in the Workbook.

Tick where you think the Sun is shining from.

WB 56

Is your shadow always the same size?

2 Measuring my shadow

Work with a partner. **Measure** how long your shadow is at different times of the day.

Write your answers in your Workbook.

Compare your shadows. Were the shadows always the same length?

Will it be a fair test if you measure a different person's shadow each time?

Hint

You should always write down the **unit** with your measurement. You can measure your shadow in centimetres (cm) or metres (m).

Extension

Find out what units other things are measured in.

WB 57

3 A shadow experiment

Draw your friend's shadow at different times during the day.

Pay attention to the position and length of the shadow.

Write the time next to each shadow that you draw.

What have you learnt about shadows? Complete the sentences in your Workbook.

WB 58

Morning at 10 a.m.

Afternoon at 3 p.m.

Why do shadows change during the day?

The Sun appears to move across the sky as the day passes.

The Sun appears low in the morning and evening but higher at midday.

Shadows change during the day because we see the Sun's position changing in the sky.

Interesting fact

Your shadow is shorter if you are near the **equator**.

Your shadow is longer if you are far from the equator.

5.2 Sundials for telling the time

In this section, I am learning:

- that we can tell the time using the Sun's position
- how a sundial works
- how to make a sundial.

Key words
sundial
time

We can use a watch to tell the time.
What else do we use to tell the time?
How did people tell the time before watches were made?

Sundials

This is a **sundial**. Long ago, people used sundials to tell the **time**.

A shadow forms on the sundial.

The Sun's position changes during the day. This makes the shadow move around the dial.

Hours are marked around the dial. The number where the shadow is tells us what time it is.

You can make your own sundial, like the one in the photo.

What time does this sundial show?

1 Making a sundial

Work in a group to make and use a sundial.

You will need a:

- piece of card
- cocktail stick
- pen
- small piece of sticky tack.

Your teacher will show you how to make the sundial.

When your sundial is ready, rest it on a flat surface. Ask your teacher to help you set it up correctly.

What time does your sundial show?

Leave your sundial outside. Go back some time later.

What time does your sundial show now?

Write two sentences about your sundial in your Workbook.

WB 59

Hint

The Sun must be shining for a sundial to work.

Extension

Find some pictures of different sundials. Copy them or stick them in your Workbook.

WB 59

2 The shadow song

Here is a song. It helps us to understand how shadows form.

Sing the song to the tune of "Old MacDonald had a farm".

Stand up and use your arms to point to the Sun and the shadow.

When the Sun is over there, my shadow is down here.

And when the Sun is over there, my shadow is down here.

Sun up there, shadow down here.

Sun there, shadow here, Sun there, shadow here.

When the Sun is over there, my shadow is down here.

Think back to what you learnt in the song.

In your Workbook, draw where you think the boy's shadow should be.

WB 60

Interesting fact

Sundials were the first type of clock. People started using sundials more than 3000 years ago. Electric clocks were invented only around 1840.

Do you think modern clocks are better than sundials? Why do you say so?

Checklist

Complete these statements on page 61 of your Workbook.

- I can say how the position and size of a shadow change during the day.
- I understand how the Sun's position causes shadows to change during the day.
- I know how a sundial works.
- I can make my own sundial.

5.3 Spinning Earth

In this section, I am learning:

- how the Earth moves on its axis
- how the movement of the Earth causes day and night.

Key words

axis
dark
day
Earth
globe
model
night
spin
torch

The Sun, the Earth and the Moon

Look at the picture.

Which part of the Earth is in daytime in the picture? Is it always daytime there?

Where is it night?

What is day and night?

Light shines from the Sun onto the **Earth**.

The Earth turns around on its **axis**.

So the Sun's light cannot reach the whole of the Earth at once. This makes daytime and night-time.

Look at the picture again. The side of the Earth facing the Sun is bright. For people living there, it is **day**.

The other side of the Earth is not getting light from the Sun now. It is **dark**. For people living there, it is **night**.

Spinning Earth

A long time ago, people thought that the Sun moved around the Earth.

They looked into the sky and saw the position of the Sun change as the day went on.

But scientists later proved that the Sun does not move around the Earth. They proved that the Earth spins around. When it spins around on its axis one side becomes day and one side becomes night.

That is why we have day and night on Earth.

The Earth has the shape of a ball. We can use a **model** to show day and night on Earth.

1

Making a model

Work with your teacher. You will need a **torch** and a ball. Work in a dark room.

One of you should hold the ball. The other then shines the torch on the ball.

What plays the part of the Sun in the model? What plays the part of the Earth in the model?

Where is it day on Earth? Where is it night on Earth?

Label the picture in your Workbook.

WB 62

2 A model for night and day

Stick figures on opposite sides of the **globe**.

Now take the globe into a dark room. Shine a torch on the globe. One figure should be in the light. The other should be in the dark.

Slowly spin the globe.

What happens to the two figures?

WB 63

Extension

What do you think will happen if the Earth stops spinning?

In your Workbook, tick which statements you think are true.

WB 63

Looking for clues

Mira's aunt lives in a country on the other side of the world. They spoke on the telephone. It was daytime for Mira, but night-time for her aunt. The girls are wondering why this was so.

Remember the model you made in Activity 1. It gives Mira, Ruth and Natalie some clues:

1 The person holding the torch stayed in one place.

2 They let the ball **spin** around.

3 They never switched off the torch.

3 **Who is right?**

Look again at page 67. In your group, talk about what Mira, Ruth and Natalie said.

Who do you think is right?

Talk about how the extra clues can help you decide.

WB 64

Checklist

Complete these statements on page 64 of your Workbook.

- I can say how the Earth moves so that we have day and night.
- I can use clues to help me work out why we have day and night.
- I can use a model to show why we have day and night.

5.4 The Moon

In this section, I am learning:

- how the Moon moves
- why we cannot always see the Moon at night.

Key words
Moon

Where is the Moon?

The **Moon** has the shape of a ball. The Moon moves around the Earth.

We can see the Moon in the sky above us. We can see the Moon because the Sun shines light onto it.

But the Moon does not always look the same to us. Sometimes we cannot see the Moon.

Look at the two photographs. How does the Moon look different in the two pictures?

How else have you seen the Moon look?

We can use a model to help us understand why the Moon does not always look the same.

1 Looking at the Moon

Listen as your teacher explains how to set up the model.

Now use the model to show the rest of the class why the Moon does not always look the same.

Who plays the part of the Earth, the Moon and the Sun in your model?

Look at the picture in your Workbook.

What will you see when the Moon is in different positions?

WB 65

Hint

You can see the Moon only when the Sun is shining onto the side of the Moon facing you.

The Moon moves around the Earth. It takes about a month to go round.

The Earth moves around the Sun. It takes a year to go round.

This means that the Moon and the Earth both move around the Sun.

Interesting fact

It takes 365¼ days for the Earth to go around the Sun once. A year is 365¼ days long.

You can use a model to show how the Earth and Moon move around the Sun together.

2 A bigger model for the Moon, Earth and Sun

Look at the children modelling the Earth, Sun and Moon.

Get into groups. Make a model like the children in the picture.

Remember that the Earth spins around itself. The Moon moves around Earth. The Sun stays still.

Tick the boxes in your Workbook to describe what your model shows about the Earth, Moon and Sun.

WB 66

Extension

You can see the Sun and the Moon in the sky. The Sun is a star.

Is the Moon also a star?

Share your ideas with your partner.

WB 67

Interesting fact

The Sun is very far from Earth. It would take a jet aeroplane about 20 years to travel as far as the Sun is from the Earth.

Checklist

Complete these statements on page 67 of your Workbook.

- I know that the Moon does not always look the same.
- I understand why the Moon does not always look the same.
- I can describe how the Moon moves.

What have I learnt?

Answer these questions on the worksheet your teacher gives you.

1 Tom, Hiro and Jason have made a chart of their playground shadows.

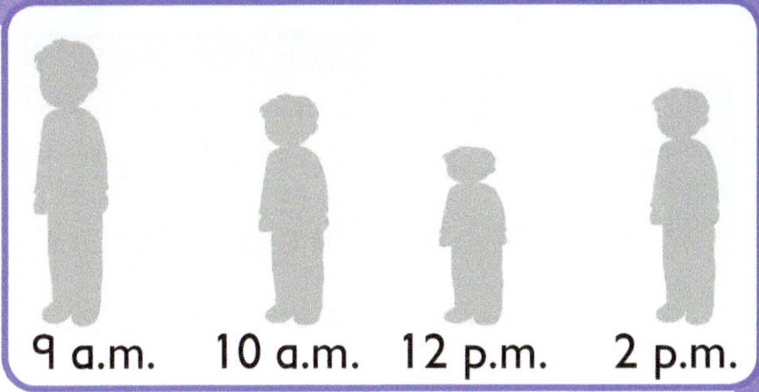

- What do you think the shadows will look like for 11 a.m., 1 p.m. and 3 p.m.?
- The boys are discussing why the 12 p.m. shadow was the shortest. Who do you think is right?

2 Look at the picture.

Which sentences are **true**? Which sentences are **false**?

- It is dark for Tina and for William.
- It is daytime for William.
- It is daytime for Tina.
- Tina can see the Moon.

6 Plants and animals around us

In this unit, I am learning:

- to compare the things living in different places
- about different habitats
- how humans can damage the environment
- how we can protect the environment
- about different types of weather
- to watch, describe and record the weather.

Key words

condition
environment
habitat
litter
measure
minibeast
weather

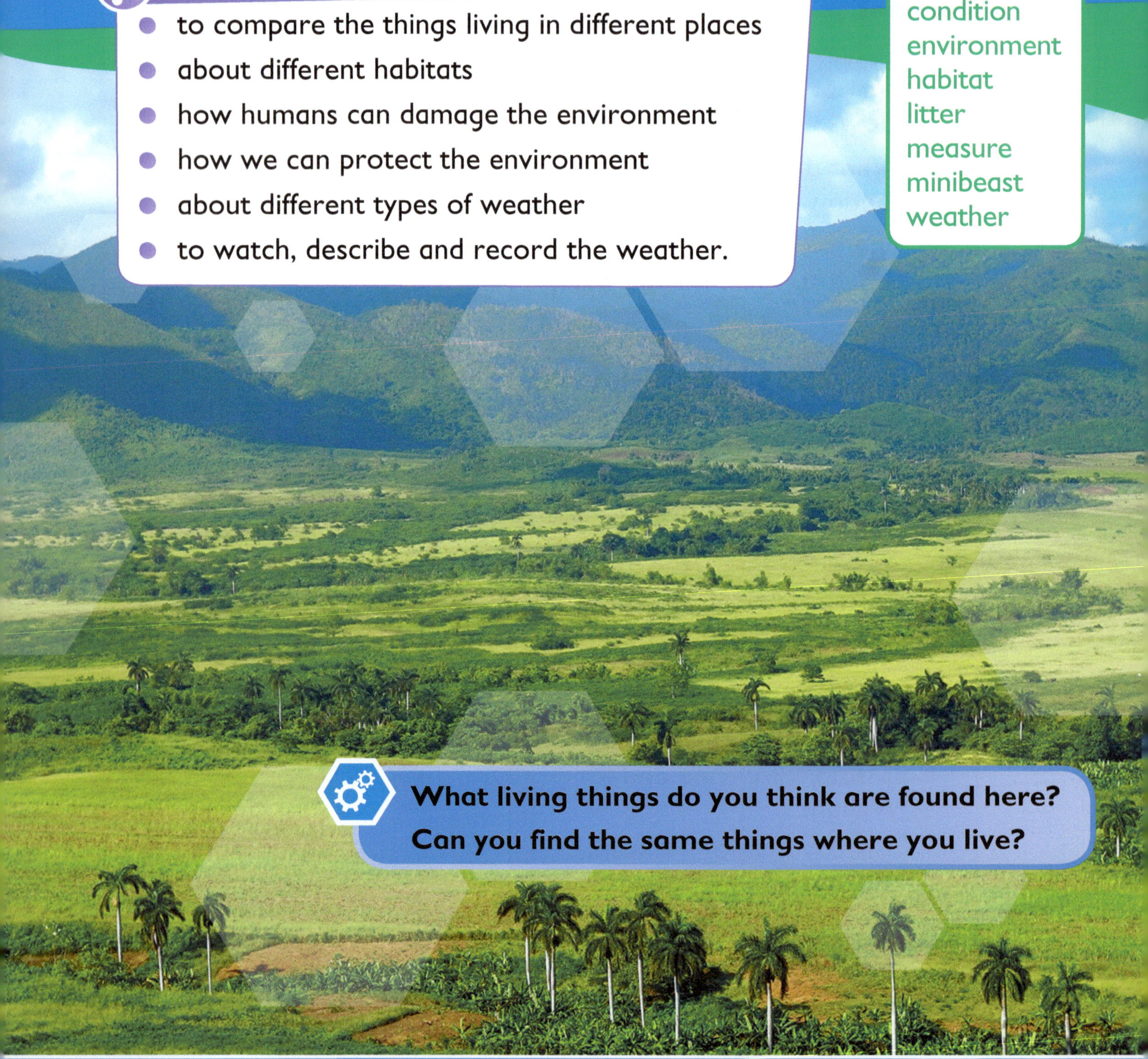

What living things do you think are found here? Can you find the same things where you live?

6.1 Habitats

In this section, I am learning:

- to compare the things living in different places
- about different habitats.

Key words

animal
condition
habitat
hot
measure
minibeast
plant
sunny
temperature
thermometer
wet

Different things live in ponds and trees.

What is it like for a plant living in a desert?

Habitats

A **habitat** is a place where things live.

A habitat gives **animals** food and shelter. A habitat gives **plants** a place to grow.

Ponds, rivers and fields are all different habitats.

There are habitats in towns as well. Parks and gardens are examples of habitats in towns.

Exploring habitats

Some habitats are **hot**. Some are cold. Some are **wet**. Some are dry.

These are all **conditions**.

An animal or plant lives in the habitat that has the right conditions.

1 Looking for different habitats

Find two different habitats outside.

Record what animals and plants you find in each habitat.

WB 70

Hint

When you explore habitats, take care not to disturb animals or plants.

Temperature is one condition in a habitat.

We can **measure** temperature with a **thermometer**. In science we measure temperature in degrees Celsius (°C).

2 What is it like in different habitats?

Go back to the different habitats you found in Activity 1.

Record some information about each one.

WB 71

Different animals and plants like different conditions to live in.

We can collect **minibeasts** to find out more about a habitat.

We can use a pooter to collect minibeasts. We can also use a pitfall trap to collect minibeasts.

Minibeasts are very small animals.

How to use a pooter

3 Collecting minibeasts

Catch some minibeasts using a pooter.

Look at the minibeasts with a magnifying glass. Record what you find.

Now get into groups. Make a pitfall trap.

Dig a hole in the ground.	Put a jar in the hole. The rim should be level with the ground.	Put some leaves or grass in the jar.	Cover the trap.

Look at the minibeasts with a magnifying glass. Record what you find.

Write about collecting minibeasts.

WB 72

Hint

After the activity, release the minibeasts back where you found them.

Extension

Where do the most minibeasts live?

Go back to the different habitats you found in Activity 1.

In each habitat, try to catch as many minibeasts as you can in a set time.

Record the numbers in a chart like the one shown to the right.

WB 73

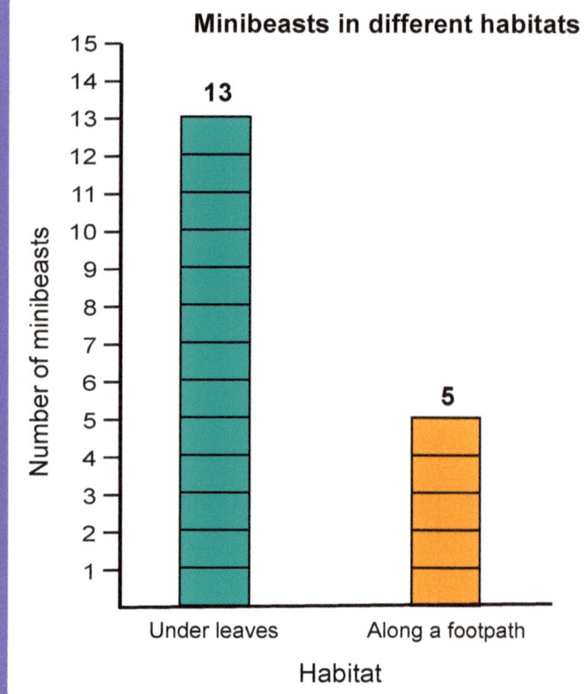

A pond is a habitat. A pond has everything that some living things need.

4 Pond dipping

You are going to look for some minibeasts in a pond.

Record what you find. Try to name some of the animals.

Did you find different minibeasts in the pond than in the other habitats?

Are the conditions in the pond the same or different compared to the other habitats?

WB 73

Hint

Do not fall into the pond!

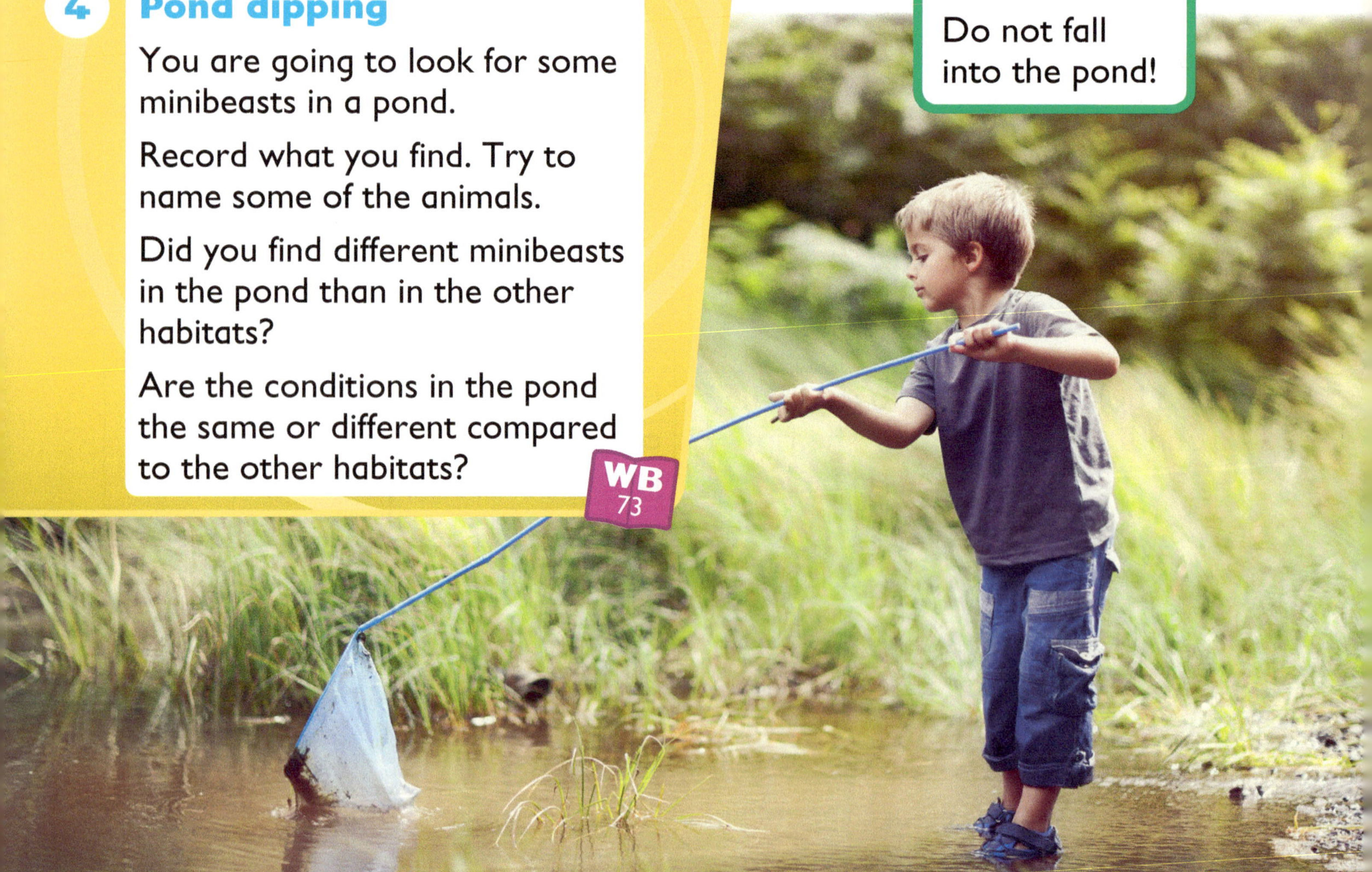

Why do we find different animals in different habitats?

Minibeasts like these live in pond water.

Remember, different habitats have different conditions.

It is shady under trees. It might also be damp there.

In a pond, it is wet. In open places, it can be **sunny** and dry.

Many minibeasts hide under things such as dead leaves on the ground.

Some minibeasts like to live in water.

Some minibeasts like dry places.

5 Matching minibeasts, habitats and conditions

Look back in your Workbook at the minibeasts you found. Think about their habitats.

Pick a minibeast from two different habitats.

Describe the conditions each one needs to live.

WB 74

Interesting fact

Coral reefs are home to many different plants, minibeasts and larger animals.

Extension

Why do so many things live on coral reefs?

WB 75

Checklist

Complete these statements on page 75 of your Workbook.

- I can name three or more different habitats.
- I can describe conditions in different habitats.
- I can say why different animals live in different habitats.
- I know of different ways to catch minibeasts.

6.2 Caring for the environment

In this section, I am learning:

- how humans can damage the environment
- how we can protect the environment.

Key words

compost
environment
litter
organic
wildlife

It is important to care for our **environment**.

Imagine that your school is building some new classrooms and a big sports centre.

What might the new development do to the habitats around the school?

How can the habitats be protected?

Environments

The word "environment" describes everything in our surroundings.

Many things affect what the environment is like, for example:

- what the weather is like
- what **wildlife** lives there
- natural things such as volcanoes
- things that humans do and build.

Some things humans do have a big effect on the environment.

If we cut down a forest to grow food, the plants and animals from the forest lose their habitats.

The animals have to find other places to live. If they cannot find another habitat, they might eventually die.

1 **What do you think?**

Imagine that your school needs to be rebuilt. The buildings are old and the roof is leaking. There are more children in the town. The community needs a bigger school.

Below is a map of the old school and the plans for the new school.

- How will building the new school affect the local environment?
- Is it possible to build the new school and keep the habitats in your local environment?

Think of ways for learners, plants and animals to all live together.

Discuss your ideas in a small group. Present your ideas to the rest of the class.

WB 76

Small changes, big differences

Everybody can do small things to help look after the environment.

Many small things add up. Picking up **litter** is a good example.

2 Litter collection

Work with a partner. Use a grabber to pick up pieces of litter. Put everything you find in a bag.

How many pieces of litter did you find around your school?

Where did you find the most litter?

What can we do so that there will be less litter?

Write your ideas in your Workbook.

WB 77

Hint

Do not pick up anything sharp. Tell your teacher about it.

There are many places for animals to live here.

Farmers can help too

Today many farmers use methods that help to care for the environment.

They do not use chemicals to kill insects that can damage their crops.

Insects are food for other animals, like birds.

We say this type of farming is **organic**.

Extension

Find out more about organic farming.

In your Workbook, draw a picture of a fruit or vegetable that comes from an organic farm.

WB 77

Interesting fact

Farmers can add **compost** to soil. This helps the crops grow. Compost is organic.

Compost is made from old plants and vegetables.

How can you and your family care for the environment?

Checklist

Complete these statements on page 78 of your Workbook.

- I can describe how humans can damage the environment.
- I can describe how humans can help to protect the environment.

6.3 Weather

In this section, I am learning:

- about different types of weather
- to watch, describe and record the weather.

Key words

cloudy
cold
frosty
icy
rainy
sunny
snowy
storm
warm
weather
windy

What do you think the weather is like in each picture? Is it warm or cold?

What clothes do people wear for this type of weather?

What other types of weather do you know of? Can you draw a picture?

When it is **rainy**, you can wear a raincoat.

Different kinds of weather

The word "**weather**" describes what is happening in the air around us.

There are very few clouds in the sky on a **sunny** day.

When we see kites fly, it is **windy**.

When it is **snowy** it is very **cold**.

On a **cloudy** day there are many clouds in the sky but little sunshine.

Weather can also be described as hot, **warm**, **frosty** or **icy**.

What do you think the weather is like when it is frosty or icy?

1 **Weather words**

Think about the types of weather.

Choose some words to describe the weather on a certain day.

Draw a picture of this type of weather.

WB 79

2 **What is the weather like today?**

Go outside, once in the morning and once in the afternoon.

What words can you use to describe the weather each time?

Did the weather change during the day?

Can you predict what the weather will be like tomorrow?

Imagine you are a TV or radio weather presenter. Record a video or podcast to describe the weather today.

WB 80

Weather symbols

A symbol is a simple picture. Symbols are useful to describe something without using words. A good symbol is simple and clear.

We can use symbols to describe the weather.

What do you think these symbols mean?

3 Making your own weather symbols

In a group, draw your own symbols for different kinds of weather on squares of paper.

Make a class display of symbols for:

- sunny weather
- windy weather
- light clouds
- big, heavy clouds
- rainy weather
- thunderstorm
- mixture of sunny and cloudy weather
- snowy weather.

Look at the symbols your classmates have drawn. Which are the clearest?

WB 80

Hint

The weather symbols used in TV weather forecasts are easy to understand. Many countries use similar symbols.

Keeping a record of the weather

We can record what the weather is like every day. When we look back later, we might see a pattern. A record can show us what the most common kind of weather is in our area.

4 My weather record

Make your own weather record in your Workbook.

Here is an example.

Day	Week 1	Week 2
Monday		
Tuesday		
Wednesday		
Thursday		
Friday		
Saturday		
Sunday		

WB 81

We can use a rain gauge to measure how much it rains.

Extension

During Activity 4, record the rainfall in your Workbook.

WB 81

5 Weather patterns

Use your weather record to make a bar chart. Here is an example.

WB 82

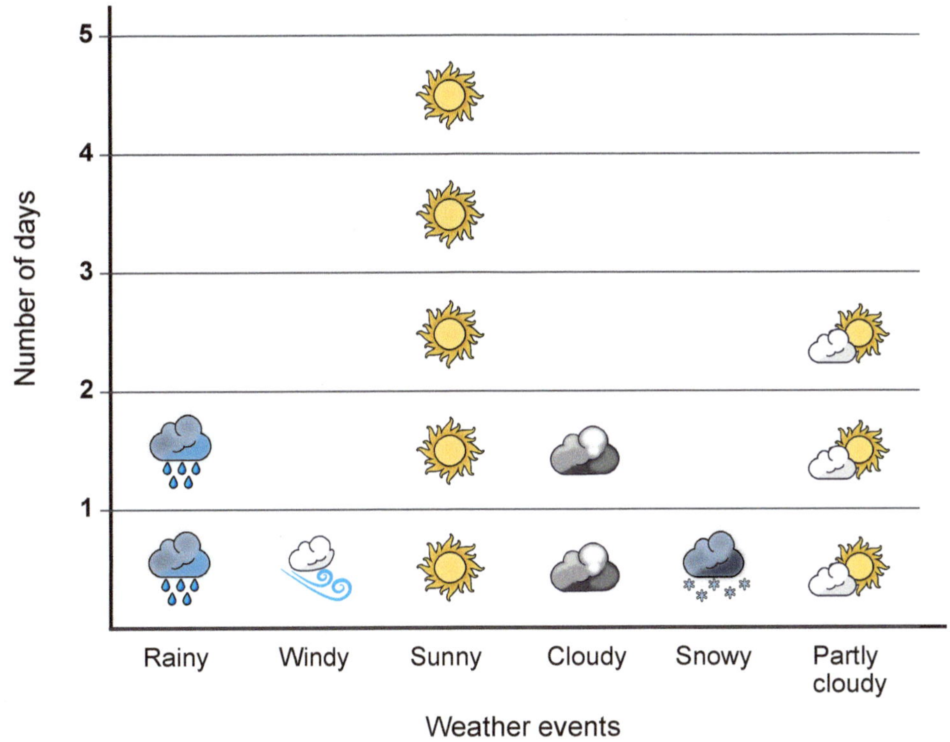

Interesting fact

A hurricane is a powerful **storm**. It brings strong winds and lots of rain. During Hurricane Irma in 2017, the wind blew at 285 kilometres per hour in some places.

Some parts of the world often have storms like this.

What harm can storms do?

Checklist

Complete these statements on page 82 of your Workbook.

- I can describe different types of weather.
- I can use symbols and charts to record the weather.

What have I learnt?

Answer these questions on the worksheet your teacher gives you.

1 Look at the picture.

- Choose the right description for each habitat labelled in the picture. You can use the same description for more than one habitat.

  wet   damp and dark   dry and light

- Match each animal to the correct habitat. An animal may live in more than one habitat.

  a bird   a swimming minibeast
  a crawling minibeast   a fish

2 Look at this picture of a town. What three things can we change to make it better for wildlife?

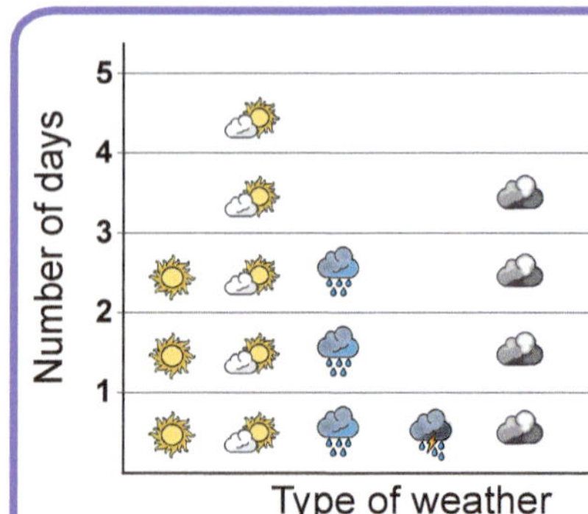

3 Look at this weather chart.

- There are five different symbols. What does each one mean?
- How many days had full sunshine?
- What type of weather was the most common?

Glossary

animal Living creature that eats and moves around on its own.
ask To make a request or seek an answer to a question.
axis A real or imaginary line around which something spins.
battery A portable source of electricity.
battery holder A box or compartment for holding a battery to keep it in position.
bend To change the shape of a material by causing it to curve.
bonding material A substance used to stick, join or fasten things together.
break To end or interrupt.
building material A material used for building structures.
bulb A device made of glass with a thin wire inside that glows when electricity passes through it.
buzzer An electrical device that makes a sound when electricity passes through it.
cell Another word for a single battery.
change To make something different.
circuit A complete path around which electricity flows.
cloudy Describing the sky when it is covered by or filled with clouds.
cold Without warmth.
compare To see how something is the same as or different from something else.
compost Pieces of old plants and vegetables that can be added to soil.
condition A characteristic of an environment.
connect To join or link two things together.
connector A device that links two objects together.
cool To remove heat.
dangerous Describing something that may cause harm.
dark Describing a place where there is little or no light.
day The time between sunrise and sunset.
depends To be affected by something else.
dissolve To mix a solid into a liquid so that the solid cannot be seen anymore.
Earth The planet where we live.
electricity What makes devices connected in a circuit work.
environment The place and conditions where a living thing lives.
equator The imaginary line that divides the Earth into a northern and a southern half.

filament A thin wire inside a light bulb that lights up when an electric current flows through it.

flow Move in a smooth, steady stream without interruption.

frosty Made of, or covered in, frost. A term often used to describe the weather when the air temperature is below 0 °C.

globe A ball showing a map of the world.

group A collection of people or things that are similar or are in one place.

guess To give or estimate an answer without knowing enough information to be certain.

habitat The place where an animal or a plant lives.

heat A form of energy that makes things hot. Heat can be measured using a thermometer.

hot Very warm.

icy Made of, or covered in, ice. A term often used to describe the weather when the air temperature is below 0 °C.

irreversible Cannot go back to how it was before.

lamp A device that gives off light.

light What makes it possible to see.

liquid Describes something that is runny or can be poured.

litter Rubbish thrown on the ground.

man-made Describes something that is made by people and does not exist in nature.

material What objects are made from.

measure To find the exact size, weight or amount of something.

melt When a solid is heated and becomes runny.

minibeast Very small animal that does not have a skeleton or bones.

model Something used to show how something else looks or works.

Moon The large ball of cold rock that moves around the Earth.

natural Describes an object or material that is found in nature and which has not been made by humans.

night The time of between sunset and sunrise when it is dark.

organic Describes material that comes from plants and animals or working in a way that protects the environment.

plant A living thing that has roots, a stem and leaves and which needs light and water to live.

position The place where something is.

predict To suggest what might happen in future based on what you already know.

rainy Describes the weather when it is raining.

results table A set of rows and columns for displaying findings or information.

reversible Can go back to how it was before.

rigid Not able to bend.

rock A big piece of hard material that comes from the Earth.

sample Small amount of a substance or material used as an example.

sand A light brown, grainy material that forms when rock is broken down.